Image-Based Visualization

Interactive Multidimensional Data Exploration

Synthesis Lectures on Visualization

David Ebert, *Purdue University*
Niklas Elmqvist, *University of Maryland*

Synthesis Lectures on Visualization publishes 50- to 100-page publications on topics pertaining to scientific visualization, information visualization, and visual analytics. Potential topics include, but are not limited to: scientific, information, and medical visualization; visual analytics, applications of visualization and analysis; mathematical foundations of visualization and analytics; interaction, cognition, and perception related to visualization and analytics; data integration, analysis, and visualization; new applications of visualization and analysis; knowledge discovery management and representation; systems, and evaluation; distributed and collaborative visualization and analysis.

Image-Based Visualization: Interactive Multidimensional Data Exploration
Christophe Hurter

ISBN: 978-3-031-01473-4   print
ISBN: 978-3-031-02601-0   ebook

DOI 10.1007/978-3-031-02601-0

A Publication in the Springer series
SYNTHESIS LECTURES ON VISUALIZATION, #4
Series Editors: David S. Ebert, Purdue University, Niklas Elmqvist, University of Maryland

Series ISSN 2159-516X Print 2159-5178 Electronic

Image-Based Visualization
Interactive Multidimensional Data Exploration

Christophe Hurter

ENAC, Ecole Nationale de l'Aviation Civile

SYNTHESIS LECTURES ON VISUALIZATION #4

ABSTRACT

Our society has entered a data-driven era, one in which not only are enormous amounts of data being generated daily but there are also growing expectations placed on the analysis of this data. Some data have become simply too large to be displayed and some have too short a lifespan to be handled properly with classical visualization or analysis methods. In order to address these issues, this book explores the potential solutions where we not only visualize data, but also allow users to be able to interact with it. Therefore, this book will focus on two main topics: large dataset visualization and interaction.

Graphic cards and their image processing power can leverage large data visualization but they can also be of great interest to support interaction. Therefore, this book will show how to take advantage of graphic card computation power with techniques called GPGPUs (general-purpose computing on graphics processing units). As specific examples, this book details GPGPU usages to produce fast enough visualization to be interactive with improved brushing techniques, fast animations between different data representations, and view simplifications (i.e. static and dynamic bundling techniques).

Since data storage and memory limitation is less and less of an issue, we will also present techniques to reduce computation time by using memory as a new tool to solve computationally challenging problems. We will investigate innovative data processing techniques: while classical algorithms are expressed in data space (e.g. computation on geographic locations), we will express them in graphic space (e.g., raster map like a screen composed of pixels). This consists of two steps: (1) a data representation is built using straightforward visualization techniques; and (2) the resulting image undergoes purely graphical transformations using image processing techniques. This type of technique is called *image-based visualization.*

The goal of this book is to explore new computing techniques using image-based techniques to provide efficient visualizations and user interfaces for the exploration of large datasets. This book concentrates on the areas of information visualization, visual analytics, computer graphics, and human-computer interaction. This book opens up a whole field of study, including the scientific validation of these techniques, their limitations, and their generalizations to different types of datasets.

KEYWORDS

human-computer interaction, interaction techniques, visualization techniques, information visualization

Contents

List of Figures

Figure Credits

CHAPTER 1

Introduction

Our society has entered a data-driven era, one in which not only are enormous amounts of data being generated daily but also growing expectations are placed on their analysis (Thomas and Cook, 2005). With the support of companies like Google,[1] big data has become a fast-emerging technology. OpenData programs, in which data are available for free, are growing in number. A number of popular websites, instead of protecting their data against "scripting," have opened access to their data through web services in exchange for pecuniary retribution (e.g., SNCF,[2] France's national railway company; IMDb movie database[3] …). Taking advantage of this, new activities are emerging, such as data journalism that consists of extracting interesting information from available data and presenting it to the public in a striking fashion (Segel and Heer, 2010) with, for instance, data videos (Amini et al., 2015). Analyzing these massive and complex datasets is essential to make new discoveries and create benefits for people, but it remains a very difficult task. Most data have become simply too large to be displayed and even the number of available pixels on a screen are not sufficient to carry every piece of information (Fekete and Plaisant, 2002). These data can also have too short a lifespan, i.e., they change too rapidly, for classical visualization or analysis methods to handle them properly and to extract relevant information in a constrained time frame.

When exploring a large dataset, two main challenges arise. The first concerns large data representation: How can these datasets be represented and how can this be done in an efficient manner? The second challenge addresses data manipulation: How can we interact effectively with them and how can this be done in a way which fosters discovery? Both interaction and representation heavily rely on algorithms: algorithms to compute and display the representation, and algorithms to transform the manipulation by the user into updates of the view and the data. Not only does the performance of these algorithms determine what representations can be used in practice, but their nature also has a strong influence on what the visualizations look like.

1.1 IMAGE-BASED VISUALIZATION

The classical way to process data is to express it in a continuous space (e.g., floating points). *Image-based visualization* investigates an alternative approach: data are expressed in the graphic space (discreet space with pixels). This approach differs from most other visualization works in that it

---

[1] http://www.google.fr
[2] http://data.sncf.com/
[3] http://www.imdb.com/

not only uses pixel-based visualization techniques, but it also performs data exploration using image-based algorithms for computer graphics. Image-based visualization explores a domain that is not just classical visualization because it relies on computer graphics, yet also not just computer graphics either because it still focuses on interaction rather than just the creation of graphics (Hurter, 2014).

As an example, the following steps show a pixel-based visualization process. These steps are inspired from the InfoVis pipeline (Card et al., 1999a) and will be detailed in the following (Figure 1.2).

1. Data can be processed with image-based algorithms in a discreet space to produce additional data dimensions (i.e., clustering). This first step is optional.

2. Data representation is constructed using readily available visualization techniques (information visualization (InfoVis), scientific visualization (SciVis), geographical visualization (GeoVis), etc.).

3. Resulting visualization undergoes purely graphical transformations using image-processing techniques (i.e., contour detection, blur filtering, hue modification, etc.).

4. Rather than only modifying the data-to-image mapping, user manipulations also modify the image processing (i.e., users manipulate the lighting of the scene to reveal visual patterns). Thanks to the image-based algorithm, these interaction are processed in real time.

As a first example, Hurter et al. (2012) investigated the mean shift algorithm (Comaniciu and Meer, 2002), a clustering computer graphic algorithm, and developed Kernel Density Estimation Edge Bundling, (KDEEB Hurter et al., 2012) a visual simplification method (image-based bundling algorithm). In Figure 1.1, a dense geographical dataset is displayed. Thanks to the image-based visualization techniques, one can see emerging patterns (right image) which were not visible in the non-processed view. The bundling technique (KDEEB) will be detailed in Chapter 4, but they rely on image-based processing with the data expressed in a raster space (pixels in a density map). The shading which shows the black and red colors is computed thanks to lighting computation (image-processing technique). Finally, the computation remains scalable thanks to the graphic card usage.

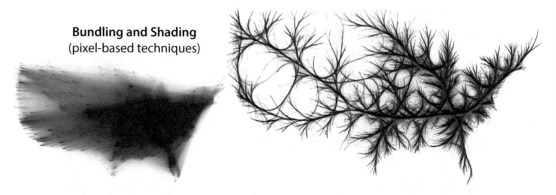

Figure 1.1: County-to-county migration flow files (http://www.census.gov/population/www/cen2000/ctytoctyflow/, the Census 2000): people who moved between counties within five years. Original data only shows the outline of the U.S. (left), bundled and shaded path (right) shows multiple information like East-West and North-South paths (Hurter et al., 2012).

In summary, image-based visualization goal is to explore new computing techniques to provide efficient interactive visualizations and user interfaces for the exploration of large datasets. Image-based visualizations are at the crossroads of information visualization, visual analytics, computer graphics and human-computer interaction.

1.1.1   DEFINITION

Image-based visualizations are specific techniques to support interactive data exploration. They combine visualization and interaction algorithms where data can be processed in a discreet space at any stage of the data-processing pipeline. Image-based visualizations can process a large dataset and usually use the graphic card parallel processing power.

1.2   IMAGE-BASED ALGORITHM OPPORTUNITIES

There are three potential benefits of image-based visualization.

1. Pixel-based algorithms can greatly benefit from the use of graphic cards and their massive memory and parallel computation power. They are highly scalable (each pixel can be used to display information) and graphic cards can easily handle a large quantity of them. In addition, classical image0processing techniques, such as sampling and filtering, can be used to construct continuous multiscale representations, which further helps scalability.

2. Image-processing fields offer many efficient algorithms that are worth applying to image-based information visualization. By synthesizing color, shading, and texture at a

pixel level, we achieve a much higher freedom in constructing a wide variety of representation that is able to depict the rich data patterns we aim to analyze.

3. The use of memory instead of computation can reduce algorithm complexity. Under a given set of restrictions, reduced complexity should reduce computation time and thus improve the ability of users to interact with complex representations. Furthermore, reduced complexity should facilitate comprehension by programmers, and thus foster maintainability, dissemination, and reuse by third parties.

1.3   THE INFORMATION VISUALIZATION PIPELINE AND ITS EXTENSION

The InfoVis Pipeline (Card et al., 1999a) introduced a model which transforms raw data into visual entities. Each stage of this transformation is customable and thus allows interactive data visualization. This conceptual model is generic enough to be applicable to any visualization system. Previous work already investigated how this pipeline could be extended to support image-based techniques. McDonnel and Elmqvist (2009) used this pipeline to support interactive InfoVis visualization. This pioneering work only investigated the very last transformation stage when the data is displayed to the user (View stage in Figure 1.2) but there is not conceptual limitation to apply image-based technique to every transformation stages. To do so, one needs to link every stage with a data storage compatible with image-based visualization. Raster map, a discreet space (i.e., pixel map or texture) is the optimal candidate (Cohen et al., 1993). More than one map can be linked to every stage and they can also be dynamically generated. Finally, these raster maps can be used for visualization but also for interaction purposes. For instance, data density maps can be computed in the View stage and stored in a raster map. These data densities can then be used as a line width to display trajectories of moving objects. On the interaction side, the user can brush in a raster map which will be used to create a new data table with selected (i.e., brushed) items.

This pipeline can be separated into three stages (Figure 1.2).

- **Data Level:** raw data are stored but can also be filtered to extract relevant data subsets.

- **Geometry Level:** data are stored into specific structured for future usages. This is especially the case when dealing with hierarchical data like graphs.

- **Image Level**: data are displayed to the user thanks to views composed of pixels.

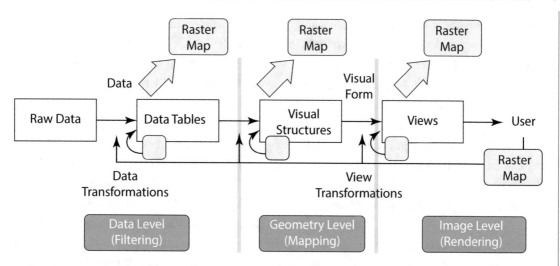

Figure 1.2: The revisited InfoVis pipeline to support image-based visualization at any stage of the data transformation (Card et al., 1999a).

1.4   GPGPU USAGES TO ADDRESS SCALABILITY ISSUES

Image-based visualizations directly take advantage of the recent progress with the graphic card programming pipe line (Jankun-Kelly et al., 2014). This section details graphic card advantages and explain how they can address scalability issues.

1.4.1   GP/GPU TECHNIQUE AND HISTORY

In this section, we focus on how graphics processors can enable scalable visualization and we give some limitations.

GPU Pipeline, Fixed vs. Programmable. The rise of special-purpose graphics hardware for the accelerated monochrome and color display of 2D/3D raster and vector graphics began during the mid to late 1970s and widespread consumer adoption, especially of hardware 3D-acceleration solutions, was obtained during the late 1990s. Such hardware was originally built around *fixed functionality pipelines* (FFPs), i.e., special purpose hardware that supports a limited and fixed set of instructions (drawing commands) to display various types of graphics primitives. Typical FFPs support operations such as geometric transformation, lighting, and rasterization, all of which are necessary for displaying and projecting 3D graphics on 2D raster displays. From the mid-2000s onward, graphics hardware manufacturers, as well as graphics API developers, gradually shifted their focus to programmable pipelines instead of FFPs. *Programmable pipelines* allow the graphics processing unit (GPU) to run proprietary code (Owens et al., 2007). Such code can be used to implement new types of drawing commands and can even be used—although initially indirectly—to perform (non-graphics-related) computational tasks on a GPU, i.e., "general purpose computation on the GPU" or GPGPU (Thompson et al., 2002). The latter is useful because of the massive parallelism offered by GPUs as well as the ease with which GPUs generally handle vector and matrix operations, a direct result of the fact that 2D/3D transformations and projections within FFPs rely heavily on vector/matrix math.

GPU Programming. APIs and pitfalls: programming each level of the graphic card pipeline can be performed through different languages, such as NVidia's Cg, Microsoft's High-Level Shading Language (HLSL), and the OpenGL shading language (GLSL). Other specialized languages exist to do specific data processing, such as CUDA and OpenCL. If we exclude specific data processing languages ("CUDA Technology," 2007) (Munshi et al., 2011) which use specific data structures, output data must be stored in image textures. Graphic cards propose massive parallel computing but some pitfalls must be avoided in order to take advantage of this worthwhile power.

- Graphic cards are optimized to compute data in parallel and therefore sequential algorithms cannot be parallelized without insuring data integrity (memory protection). Reading and writing graphics memory is not possible at the same time; this avoids memory corruption (one process reading at the same time as another is updating the information).

- Synchronization features, such as mutex or memory protection (atomic functions), must be avoided as much as possible. Specific computation techniques can be applied such as MapReduce, a programming model for processing large data sets with a parallel, distributed algorithm on a cluster (He et al., 2008).

- Bottlenecks exist within the GPU processing, especially when transferring data between the CPU and GPU. When this occurs, the graphic card needs to wait until every process has ended, and then start the memory transfer—a dramatically slower process. Memory transfer between the GPU and CPU must be limited as much as possible.

- Many other pitfalls must be taken into account regarding each language, such as texture coordinates that differ between OpenGL and DirectX, debugging issues, and graphic card crashes that hinder the development process.

1.4.2   IMAGE-BASED AND THE GRAPHIC CARD

The image-based approach takes advantage of changes in the bottlenecks of computer graphics: since data storage and memory limitation is becoming less and less of an issue (Sutherland, 2012), we can plan to reduce computation time by using memory as a new tool to solve computationally challenging problems. Furthermore, even if graphic cards were initially developed to produce 2D/3D views close to photo-realistic images, their power has also been used to perform parallel computations (so-called GPGPU techniques) (Owens et al., 2007).

The following list provides instances of image-based techniques with the graphic card usage to facilitate multivariate data exploration; they use multi-pass read-write cycles, minimize CPU-GPU memory transfer, and accommodate variation in graphical hardware. They can be classified into three groups: rendering, computation, and interaction.

- **Rendering:** Graphic cards can render numerous items on the screen and thus can display large datasets. In the following examples, GPUs are used to display data and to perform image-based rendering techniques. Auber developed Tulip (Auber, 2004), an information visualization framework dedicated to the analysis and visualization of relational data. This software uses GP-GPU techniques to render large multivariate graphs. McDonnel and Elmqvist (2009) developed a framework and an application using shaders to display multivariate data based on the dataflow model with a final image-based stage. In this final step, the multivariate data of the visualization are sampled in the resolution of the current view. A more specific rendering technique is used by Holten (2006) to improve edge visualization by an interesting variation on standard alpha blending, i.e., how color transparency is combined. Scheepens et al. (2011) used the GPU to compute density maps and then apply shading techniques to emphasize multivariate data on the density map of moving vessels.

- **Computation:** Graphic cards can perform fast and parallel data processing, and can be used to process information at the data level. FromDaDy, From Data to DisplaY (Hurter et al., 2009b) uses the GPU for interactive exploration of multivariate relational data. Given a spatial embedding of the data, in terms of a scatter plot or graph layout, the Mole View (Hurter et al., 2011) uses a semantic lens which selects a specific spatial and attribute-related data range. The lens keeps the selected data in focus unchanged and continuously deforms the data out of the selection range in order to maintain the context around the focus. Animation is also performed between the bundled and the unbundled layout of a graph. Kernel Density Edge Bundling (KDEB) (Hurter et al., 2012) computes bundled layouts of general graphs. This technique is also applied on dynamic graphs (Hurter et al., 2013a). Other GPU bundling techniques also exist, i.e., winding roads uses a voronoï diagram to compute Graph bundling and its density (Lambert et al., 2010a). Finally, the GPU has been used directly for graph layout as well (Frishman and Tal, 2007).

- **Interaction:** Interaction with the data is an important manipulation paradigm to perform data exploration. Graphic cards can be used to provide tools to help users to interact with large datasets. Rolling the dice (Elmqvist et al., 2008) helps the user to define the appropriate displayed variables with a smooth animation when changing visual configuration; Graphdice (Bezerianos et al., 2010) uses the same paradigms but with graphs. FromDaDy (Hurter et al., 2009b) uses related animation with GP-GPU techniques. In order to address the dataset size issue, FromDaDy loads the whole dataset within the graphic card so that no memory transfer is needed when changing visual configuration. This helps to improve interaction with fast and continuous animations. Furthermore, a GP-GPU technique is implemented to support brushing and data manipulation across multiple views. The user can then brush trajectories, and with a pick and drop operation he or she can spread the brushed information across views. Thanks to the usage of graphical textures as memory storage, this interaction can be repeated to extract a set of relevant data, thus formulating complex queries.

Conceptually, image-based visualization takes its roots in the following pioneering works:

- texturing as a fundamental primitive (Cohen et al., 1993);

- cushion Treemaps (Van Wijk and van de Wetering, 1999);

- dense-pixel visualizations (Fekete and Plaisant, 2002) which use every available pixel of an image to carry information; and

- GPU usages in InfoVis (McDonnel and Elmqvist, 2009).

1.5   DATA TYPES

Many types of datasets do exist, each of them with specific constraints and analysis purposes. Image-based visualization can be applicable to many data types with various challenges. There is no restriction for using such visualization techniques with multivariate data, but in the following, we list relevant data types and discuss how image-based visualization can be of interest to visualize and interact with them.

1.5.1   TIME-DEPENDENT DATA

Analyzing and understanding time-dependent data pose non-trivial challenges to information visualization. Since such datasets evolve over time, they can be several orders of magnitude larger than the same static datasets (i.e., only one temporal instance). This can be an issue, especially if their analysis must be performed in a constrained time frame which underlines the importance of relying on efficient interactions with multiple objects and fast algorithms. In addition, while patterns of interest in static data can be naturally depicted by specific representations, we do not yet know the best way to visualize dynamic patterns, which are inherent to time-dependent data. While there are many solutions for displaying patterns of interest in static data with still visualizations, few previous works have addressed the issue of dynamic patterns (von Landesberger et al., 2011). Since image-based visualization techniques are scalable, they represent a great opportunity for such data types.

1.5.2   MOVEMENT DATA

Movement data, which are multidimensional time-dependent data, describe changes in the spatial positions of discrete mobile objects. Automatically collected movement data (e.g., GPS, RFID, radars, and others) are semantically poor as they basically consist of object identifiers, coordinates in space, and time stamps. Despite this, valuable information about the objects and their movement behavior, as well as about the space and time in which they move, can be gained from movement data by means of analysis. Such data usually represent a large amount of records which hinder their visualization (not enough pixel to display all of them) and analysis (analysis must be performed in real time to foster information retrieval). Image-based techniques have already shown their efficiency with bundling techniques (Hurter et al., 2012) but many research opportunities remain.

1.5.3   GRAPH DATA

The Infovis Pipeline and its three-stage structure (Figure 1.2) has been the core of a tutorial (Schulz and Hurter, 2013) to present graph simplification techniques. This tutorial gives solutions to address dense graph visualization and exploration. Dense graph visualization and the resulting "hairball" suffer from cluttering and overplotting to an extreme that renders it unusable for any practical purposes. Since researchers have had this experience for decades, various approaches have been developed on all stages of the visualization pipeline to alleviate this problem. Image-based visualization can be directly applicable to support graph exploration on various tasks: filtering, clustering, aggregation, simplification, patterns finding, etc. Graph exploration can also be challenging with dynamic graphs where their structure and their content evolve over time (Hurter et al., 2013a).

1.6   BOOK ROADMAP

This first chapter introduced image-based visualization and highlighted how such technique can leverage multivariate data exploration. Thanks to the increasing power of graphic cards these techniques can unleash all of their power to interactively process data. As such, the user will have new customable tools within interactive visualization to support large data exploration. Chapter 2 will show how this concept emerged from previous research. In Chapter 3 one of the main uses of image-based visualization will be detailed with interactive density map computation and visualization. In Chapter 5, we will present how animations and distortions are also applicable to such techniques. Finally, Chapter 6 details future usages and challenges which remain to be addressed.

CHAPTER 2

Motivating Example

This chapter presents the journey that motivated the investigation of image-based visualizations. This journey started with a Ph.D. in visualization characterization (Hurter, 2010). Although at first this might not seem to correlate with interactive data exploration, these two topics in fact share the same backbones with the data transformation pipeline (Card et al., 1999b). The initial goal of this Ph.D. was to find and develop efficient tools to describe (i.e., characterize) visualizations. The characterization of visualization is a wide and complex topic in which many research remains to be done. Nevertheless, this characterization topic shows the path to better understand how information can be presented to a user, and how data can be transformed through many separated steps in order to end up visible on a raster map composed of pixels on a screen. One of the first issues to better understand this transformation process is to display more and more data on a screen. This is a reason why image-based techniques had to be deployed to address visualization and interaction issues with large datasets. This chapter will detail this journey starting with the development of tools devoted to designer purposes and transposed to display large quantities of data.

2.1 VISUALIZATION EVALUATION

The evaluation of visualizations is a complex topic which is often based on the completion time and error measurement to perform a task. Since users are involved in the evaluation process, this method is time consuming and requires numerous users to yield reliable results. Some methods exist to assess visualizations before user tests but they only concern the effectiveness of interaction. These methods rely on models of the system and they have proved to be accurate and efficient when designing new interfaces. For example, KeyStroke (Card et al., 1983) and CIS (Appert et al., 2005) are *predictive* models that help compute a measurement of expected effectiveness, and enable quantitative comparison between interaction techniques. If methods to assess interactive systems do exist, very few can assess visualization before user tests.

The goal it to go beyond *time* and *error* evaluation and propose an assessment of the *bandwidth* of available information in a visualization. Therefore, analyzing visualizations to extract relevant characterization dimensions is highly relevant to this task. The objective is to perform an accurate visualization evaluation and to answer these questions: What is the visible information? What are the phenomena or mechanisms that make them visible? To characterize visualization, one faces the following additional questions.

- How can the relevant characterization dimensions for the description be found (the content of the description)?

- How can an accurate and exhaustive description of a visualization be formatted?

- How can the characterization be represented to enable comparisons?

Previous works use the data transformation pipeline to find relevant characterization dimensions of a visualization (Card and Mackinlay, 1996). This pipeline model uses raw data as an input and alters them with transformation functions to produce visual entities as an output. Thus, the characterization of visualization consists of describing the transformation functions. However, this method is not sufficient to fully describe visualization, especially with a specific class of design that uses emerging information (Hurter and Conversy, 2008). Basically, emerging information is perceived by users without being transformed by the pipeline model functions.

The first step of this work was to gather enough examples of visualization to cover the largest design space. Then one can apply available characterization models to assess if they were suitable for the activity to be supported and, if the need arose, to improve them. This characterization had to be done with objective and formal assessments. The designer should have been able to use this characterization to list the available information, to compare the differences between views, to understand them, and to communicate with accurate statements (Hurter, 2010).

2.2   APPLICATION DOMAIN

In order to benefit from concrete cases, we used the Air Traffic Control (ATC) application domain. ATC activities employ two kinds of visualization systems: real-time traffic views, which are used by Air Traffic Control Officers (ATCOs) to monitor aircraft positions, and data analysis systems, used by experts to analyze past traffic recording (e.g., conflict analysis or traffic workload). Both types of systems employ complex and dynamic visualizations, displaying hundreds of data items that must be understood with minimum cognitive workload.

As traffic increases together with safety concerns, ATC systems need to display more data items with at least the same efficiency as existing visualizations. However, the lack of efficient methods to analyze and understand why a particular visualization is effective spoils the design process. Since designers have difficulty analyzing previous systems, they are not able to improve them successfully or communicate accurately about design concerns. Visualization analysis can be performed by *characterizing* them. In the InfoVis field, existing characterizing tools are based on the dataflow model (Card et al., 1999b) that takes raw data as input and produces visualizations with transformation functions. Even if this model is able to build most of the existing classes of visualization, we show in the following that it is not able to characterize them fully, especially in the case of ecological designs that allow emerging information.

2.2.1   INSTANCE OF DESIGN EVALUATION: THE RADAR COMET

The main task for an ATCO is to maintain a safe distance between aircraft. To be compatible with this task, the process of retrieving and analyzing information must not be cognitively costly. The precise analysis of visualization is especially useful in this field to list the visually available information, and to forecast the resulting cognitive workload.

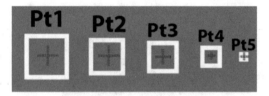

Figure 2.1: The design of the radar comet used by Air Traffic Controllers.

ODS Coded Information	Visual Code
Aircraft position	Position
ageng of each position	Size
Aircraft speed	Size (comet length)
Aircraft tendency (left, right)	Comet curvature
Aircraft acceleration	Regular/irregular point spacing
Aircraft entity	Gestalt (proximity and size)

Table 2.1: Information coded with a radar comet

As an example, in the radar view, comets display the position of the aircraft. The design of the comet is constructed with squares (Figure 2.1), whose sizes vary with the proximity in time of the aircraft's position: the biggest square displays the latest position of the aircraft, whereas the smallest square displays the least recent aircraft position. The positions of the aircraft merge through the effect of Gestalt continuity, in which a line emerges with its particular characteristics (curve, regularity of the texture formed by the points, etc.); thus, this design codes a large amount of information (Table 2.1).

Before describing the comet design fully, it is interesting to understand where the design comes from. In fact, the visual features of the comet were first used in the early 17th century by Edmond Halley (Thrower, 1969). In Figure 2.3), the comet helps to understand the trade wind direction with a thicker stroke representing the head of the comet.

The radar comet, used by ATCs, has the same properties as the one introduced by Halley, but this design was created with technological considerations in mind. Early radar screens used the phosphorescent screen effect to display the position of the aircraft. Between two radar updates, the previous position of an aircraft was still visible, with a lower intensity. This effect is called the *remanence* effect. Thus, the radar plot has a longer lifetime than the radar period (Figure 2.2). The resulting shape codes the direction of the aircraft, its speed, acceleration, and tendency (the aircraft is tending to turn right or left). For instance, Figure 2.1 displays an aircraft that is turning to the right and has accelerated (the non-constant spacing indicates the increase in aircraft speed). With technological improvements, remanence disappears, together with the additional information it provides. Designers and users felt the need to keep the remanence effect, and emulated it.

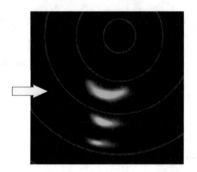 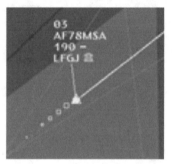

Figure 2.2: The radar design with old radar technology (left), and a modern radar screen (right).

A deeper analysis of the comet design allows us to understand that the user perceives an emerging shape: the regular layout of squares, and the regular decrease in size, configure in a line with the Gestalt effect (Koffa, 1963). Not only does a new visual entity emerge, but its own graphical properties (length and curvature) emerge as well. The graphical properties encode additional information (speed and tendency, respectively). Furthermore, this line is able to "resist" comet overlapping: the user can still understand which comet is which, despite tangling.

The design of the comet and its associated information are summarized in Table 2.2. All emerging information is due to the comet design that uses remanence: several instances of the same object at different times.

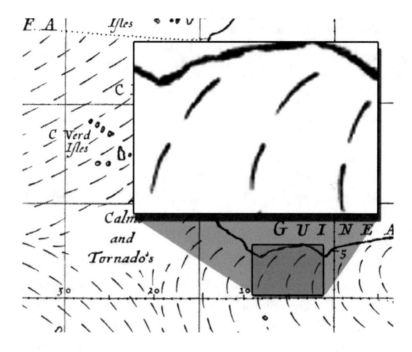

Figure 2.3: Halley drawing (1686) of the trade wind.

Table 2.2: Radar comet characterization

Transformation Function	Source of Emerging Information	Emerging Data
Point (latitude, longitude)	comet length	speed
	curvature	tendency
Size(time)	square relative positions	direction
	square spacing	acceleration

It can be noted that this design exhibits a large amount of emerging information. We can thus say that the radar comet is efficient in this respect. This is the reason why this kind of design is widely used in other visualizations (mnemonic rendering (Bezerianos et al., 2006); phosphor (Baudisch et al., 2006)).

2.3   THE CARD AND MACKINLAY MODEL IMPROVEMENTS

Card and Mackinlay (1996) attempted to establish comparison criteria of visualizations. They proposed a table for each transformation function (Table 2.3). The C&M table is completed with the notations in Table 2.4.

Table 2.3: Card and Mackinlay representation model											
			Automatic Perception								Controlled Perception
Name	D	F	D'	X	Y	Z	T	R	—	[]	CP

Table 2.4: Card and Mackinlay model notation			
S	Size	Lon, Lat	Longitude, Latitude
Sh	Shape	P	Point
f	Function	O	Orientation
N, O, Q	Nominal, Ordered, Quantitative		

The horizontal rows correspond to the input data. The column D and D' indicate the type of data (Nominal, Ordered, or Quantitative). F is a function or a filter which transforms or creates a subset of D. Columns X, Y, Z, T, R, -, [] are derived from the visual variables of Bertin (1983). The image has four dimensions: X, Y, Z, and time, T. R corresponds to the retinal perception that describes the method employed to represent information visually (color, form, size, etc.). The bonds between the graphic entities are noted with "-," and the concept of encapsulation is symbolized by "[]." Finally, a distinction is made if the representation of the data is treated by our perceptive system in an automatic or controlled way.

One can apply the Card and Mackinlay model (1996) to different kinds of ATC visualization (Hurter and Conversy, 2008; Hurter et al., 2008). When studying the radar comet, the concept of current time was introduced (Tcur: the time when the image is displayed). The size of the square is linearly proportional to current time with respect to its aging. The grey row and column are two additional items from the original C&M model (Table 2.5).

Table 2.5: C&M radar comet characterization											
Name	D	F	D'	X	Y	Z	T	R	-	[]	CP
X	QLon	f	QLon	P							
Y	QLat	f	QLat		P						
T	Q	f(Tcur)	Q					S			

However, the characterization cannot integrate controllers' perception of the aircraft's recent evolution (speed, evolution of speed, and direction). For instance, in Figure 2.1, the shape of the comet indicates that the plane has turned 90° to the right and that it has accelerated (the variation of the dot spacing). These data are important for the ATCO. The comet curvature and the aircraft acceleration cannot be characterized with this model because they constitute emerging information (there is no raw data called "curvature" to produce a curving comet with the dataflow model). In Table 2.1, italic script represents emerging information.

Whereas Card and Mackinlay (1996) depicted some InfoVis visualizations without explicitly demonstrating how to use their model, we have shown the practical effectiveness of the C&M model when characterizing the radar comet (Hurter and Conversy, 2008). Although the C&M tables make visualization amenable to analysis as well as to comparison, this model does not allow essential information to be highlighted for designers, and does not allow any exhaustive comparison of different designs. We extended this model with the characterization of emerging data. The emerging process stems from the embedded time in the radar plot positions. The time can be easily derived into speed and acceleration. We communicated about this work in a workshop (Hurter et al., 2009a) and extended it with the analysis of the visual scan path the user has to perform to retrieve a given information (Conversy et al., 2011).

2.4   CHARACTERIZATION OR DATA EXPLORATION TOOL

To support the characterization of visualizations, a simple software based on the dataflow model was developed. It was based on the following assumption: if this software can produce a valid visualization thanks to its description, then this description is also valid and correctly characterizes this visualization. This software found its inspiration in the previous work of Bertin (1983a) with a graphical description, Baudel (2004) and Wilkinson et al. (2005) with a description close to a programmable language, and finally with the C&M characterization table (Card and Mackinlay, 1996).

This prototype is called DataScreenBinder since it takes as an input a data table and with connected lines then binds fields of the dataset to visual variables (Bertin, 1983a). This prototype managed to replicate and thus provide a potential characterization of the radar screen used by ATCs (Figure 2.4).

Even if this prototype is able to duplicate existing visualization, the produced characterization was not fully suitable to support their detailed comparison. Only a visual comparison between connected lines and the data field names can be performed which is too limited.

Nevertheless, this prototype is a better fit for some other purposes such as data exploration with different visual mapping. For instance, the same dataset (Figure 2.4) can be visualized with a circular layout of aircraft speed (Figure 2.5). Such visualization shows that aircraft flying at high altitudes (large blue dots) also have fast speeds (close to the border of the circular shape).

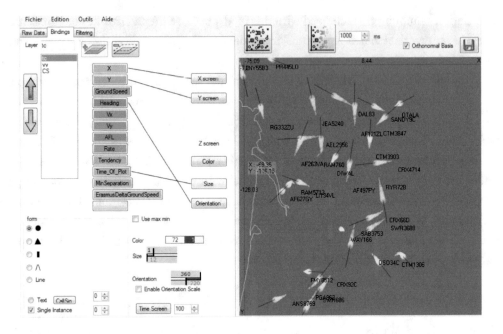

Figure 2.4: The visual caracterization of the radar screen for ATCs.

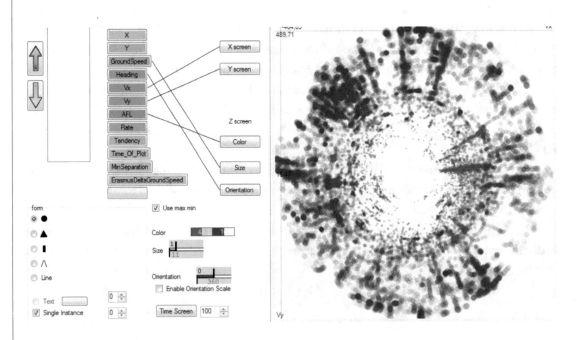

Figure 2.5: DataScreenBinder and the visualization of aircraft speeds in a circular layout.

This prototype has been intensively extended with interactive techniques like pan, zoom, dynamic filtering, selection layers, etc. This prototype was written in C# and used the GDI graphic library, which hindered the visualization of large datasets (up to 10,000 displayed dots). Therefore I started to use OpenGL/DirectX to support large dataset visualization. This worked very well until the implementation of animation and brushing techniques. To support such tasks with an interactive frame rate, this prototype had to investigate the GPGPU technique (Owens et al., 2008) and thus it ended up with the development of the software FromDaDy (Hurter et al., 2009b) with the help of detailed features (Tissoires, 2011).

2.5   FROMDADY: FROM DATA TO DISPLAY

Thanks to the first investigation with DataScreenBinder, FromDaDy (*From Data to Display;* Hurter et al., 2009b) was developed. This multidimensional data exploration is based on scatterplots, brushing, pick and drop, juxtaposed views, rapid visual design (Figure 2.6) and smooth transition between different visualization layouts (Figure 2.7). Users can organize the workspace composed of multiple juxtaposed views. They can define the visual configuration of the views by connecting data dimensions from the dataset to Bertin's visual variables. One can brush trajectories and, with a pick and drop operation, spread them across views. One can repeat these interactions until a set of relevant data has been extracted, thereby formulating complex queries.

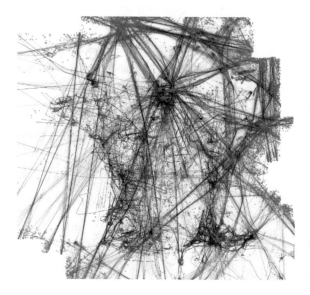

Figure 2.6: One day of recraded aicraft trajectory over France.

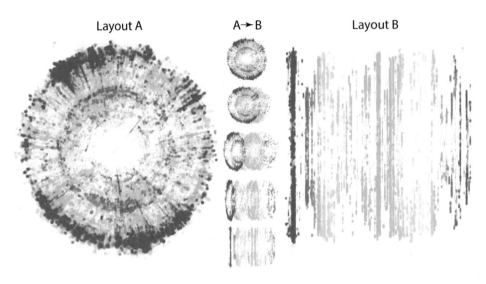

Figure 2.7: FromDaDy with two layouts and its animation.

FromDady eases Boolean operations since any operation of the interaction paradigm (brushing, picking, and dropping) implicitly performs such operations. Boolean operations are usually cumbersome to produce, even with an astute interface, as results are difficult to foresee (Young and Shneiderman, 1993). The following example illustrates the union, intersection, and negation Boolean operations. With these three basic operations the user can perform any kind of Boolean operation: AND, OR, NOT, XOR, etc. In Figure 2.8, the user wants to select trajectories that pass through region A or region B. He or she just has to brush the two desired regions and Pick/Drop the selected tracks into a new view. The resulting view contains his or her query, and the previous one contains the negation of the query.

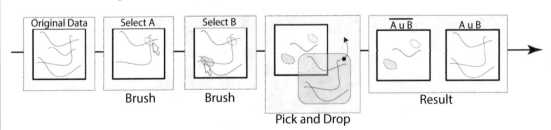

Figure 2.8: Union Boolean operation.

2.6   CONCLUSION

This chapter summarized the work to support the characterization of visualizations with ad hoc methods to depict the *bandwidth* of available information in designs. Thanks to a table as a representation for the description, one can describe designs that use emerging information.

Starting from the characterization of visualization and the need to better understand the transformation process from raw data to view, this journey led to the investigation of how large data can be transformed onto the screen composed of pixels. This was the first step to investigating pixel-based visualization. With serendipity, the characterization of visualization led toward the usage of image-based techniques with the brushing within FromDaDy. The ATC application domain was also of a great help to provide concrete application domains with specific requirements and needs. FromDaDy was the first tool developed to explore ATC data, but it is not limited to such a dataset. In fact, an increasing number of researchers and ATC practitioners are using it which leads to numerous improvements and new open research questions. Graphic card usage was also of a great interest to address scalability issues with fascinating and promising emerging technologies like OpenCL/CUDA. The chapters that follow will provide additional instances and details in which pixel-based visualization leverages interactive large dataset exploration.

CHAPTER 3

Data Density Maps

The use of the popular scatterplot method (Cleveland, 1993) is not sufficient to display all information because a great deal of overlapping occurs. When transforming data to graphical marks, a regular visualization system draws each graphical mark independently from the others: if a mark to be drawn happens to be at the same position as previously drawn marks, the system replaces (or merges using color blending) the pixels in the resulting image. The standard visualization of this pixel accumulation process is not sufficient to accurately assess their density. For instance, Figure 3.1 (left) shows one day of recorded aircraft trajectories over France with the standard color blending method. Figure 3.1 (right) shows the same dataset with a 3D and shaded density map and one can easily perceive that the data density is drastically higher over the Paris area which is not that obvious with the standard view.

Figure 3.1: Day aircraft trajectory over France (left), and 3D density map (right).

This work shows the investigations with density computation algorithm and hardware-accelerated extension of FromDaDy (Hurter et al., 2009b) to support the exploration of aircraft trajectories (Hurter et al., 2010) with the Kernel Density Estimation (Silverman, 1986).

3.1   KERNEL DENSITY ESTIMATION: AN IMAGE-BASED TECHNIQUE

Kernel Density Estimation (KDE; Silverman, 1986) is a mathematical method that computes density by a convolution of a kernel K (Figure 3.2) with data points. This method produces a smooth data aggregation which also reduces data sampling artifacts and is suitable for showing an overview of amounts of data.

Given a graph $G = \{e_i\}_{1 < i < N}$ consisting of edges $e_i \in \mathbb{R}^2$ and a point $x \in G$, we can estimate the local spatial density ρ of points x using the KDE:

$$\rho(x) = \sum_{i=1}^{N} \int_{y \in e_i} K\left(\frac{x-y}{h}\right) ;$$

Where $K: \mathbb{R}^2 \to \mathbb{R}^+$ is the so-called density of bandwidth $h > 0$. Typical kernel choices are Gaussian and Epanechnikov (quadratic) functions (Figure 3.2). ρ can be computed by convolving G with K, or building an accumulation map of K over G.

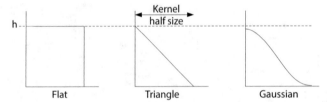

Figure 3.2: Kernel profiles.

The density ρ can be visualized as a 2D height field by a straightforward color map, contour plot, or terrain map (Figure 3.1). Landscape visualization with hills and valleys have been shown to be easy to interpret (Wise et al., 1995). For quantitative analysis, a contour plot is preferred over a colormap, since value estimation by colors is perceptually difficult. Since contour plots only use isolines, color can be used for other purposes. In 2D, the density plot becomes visually more detailed by using shading and can be enriched to a contour map (van Wijk and Telea, 2001). KDE maps can be interactively explored and modified (Van Liere and Leeuw, 2003). The KDE algorithm has also been used to investigate objects movements (Willems et al., 2009; Scheepens et al., 2011).

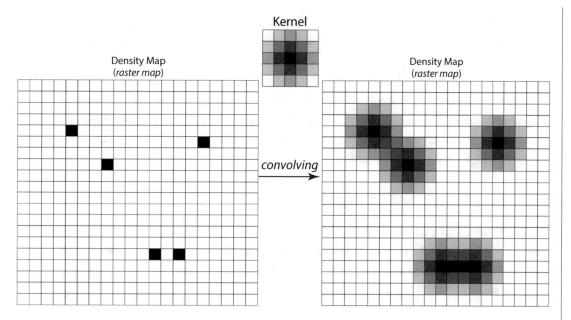

Figure 3.3: KDE convolving principle with a raster map.

Figure 3.3 shows a simple principle to compute the local spatial density of points with a kernel and a density raster map. The kernel map is applied (convolved) at every point location. This technique can be considered as an image-based algorithm and produces a grid with a smooth transition between the cells (Figure 3.1).

3.1.1   GPU IMPLEMENTATION

The initial version of FromDaDy computed the density map and stored it in an off-screen buffer with additive blending enabled. To achieve this, fromDaDy used a 32-bit floating color texture. More recent versions use OpenCL with the full support of the convolution process with tabular data. To ensure data integrity with the multi-threaded data processing, FromDaDy uses atomic functions (Shreiner and Group, 2009). The density map computation paradigm with atomic function is far from being the fastest one, but this ensures accurate results with reasonable computation time. This GPU implementation allows interactive manipulation and visualization of the density map. The computation time varies with the number of points in the data set and the kernel size. As an example, the frame rate is around 10 frames per second with 400,000 points and a kernel point size of 20 pixels and a Nvidia GTX 275 graphic card.

This GPU implementation is the cornerstone of the image-based visualization of density maps. Without this implementation, the computation time would be prohibitive especially with a large density map (more than 400x400 pixels size). This is especially important since the density computation results can drastically vary with the kernel size and shape. Many trials can be requested before achieving a suitable density map computation with interesting emerging patterns. Therefore, this exploration process requires an interactive response time to compute these density maps. Thanks to the graphic card computation power, this image-based density computation can support interactive data exploration. Even if graphic card power never stops improving, density computation remains challenging with a large dataset or big kernel size. Improvements in this area are possible by, for instance, using a better algorithm such as Divide and Conquer to implement an efficient parallel density map computation.

3.2   INTERACTION TECHNIQUES

The following describes a set of interaction and visualization techniques with density maps to perform interactive data exploration. Thanks to a GPU implementation the users can interact in real time with the density map and the process is divided into three steps.

- Users can choose which data dimensions to accumulate and can adjust the kernel size.

- Users can "brush, pick, and drop" data to remove them from, or add them, to the density map.

- Users can explicitly choose to use the computed density values with one of the available design customizations (color, size, or position).

3.2.1   BRUSHING TECHNIQUE

The brushing technique with numerous points is technologically challenging. Therefore, one has to take full advantage of modern graphic card features. FromDaDy uses a fragment shader and the render-to-texture technique (Harris, 2005). Each trajectory has a unique identifier. A texture (stored in the graphic card) contains the Boolean selection value of each trajectory, false by default. When the trajectory is brushed its value is set to true. The graphic card uses parallel rendering which prevents reading and writing in the same texture in a single pass. Therefore we used a two-step rendering process (Figure 3.4): (1) we test the intersection of the brushing shape and the point to be rendered to update the selected identifier texture and (2) we draw all the points with their corresponding selected attribute (gray color if selected, visual configuration color otherwise). This technique illustrates the very first usage of an image-based algorithm to outperform brushing technique.

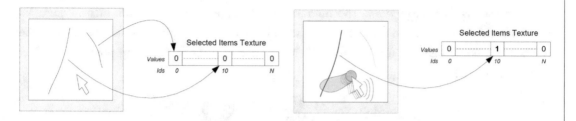

Figure 3.4: GPU implementation of the brushing technique.

3.2.2   BRUSHING TECHNIQUE WITH DENSITY MAPS

Originally, FromDaDy supported the brushing of trajectories with their spreading across views. This interaction helps to select an entire trajectory with the brushing of only few points, but in certain cases, the data exploration requires only parts of trajectories. We added the brushing of points, which allows the selection and manipulation of points. The user uses a size configurable round shape to brush the view to selected trajectories or points (Figure 3.5).

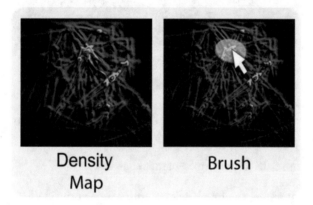

Figure 3.5: Brushing over a density map.

Thanks to the brushing technique, the user can select and highlight parts of the displayed data. By pressing the space bar, the user can extract previously selected data and attach them to the mouse cursor. By default, the selected data are *picked*: they are removed from the view, and appear in a "fly-over" view. When the user presses the space bar for the second time, a *drop* occurs in another view under the cursor. Although it resembles a regular drag'n'drop operation, FromDaDy uses the term "pick'n'drop" (Rekimoto, 1997) in the sense that data is removed from the previous view and is attached to the mouse even if the space bar is released (Hurter et al., 2009b). Finally, the most appropriated term is "Pick and Spread" which fully corresponds to the FromDaDy paradigm: spreading data across views. This interaction also applies to accumulation maps. Figure 3.6 shows the difference between the point and the trajectory mode. With the point mode, only the brushed points are selected and isolated. With the trajectory mode, brushing points also select their entire trajectory.

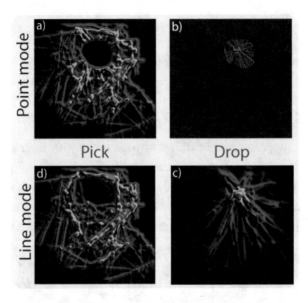

Figure 3.6: Pick and spread of points or trajectories on a density map.

The pick and spread of brushed data, from the accumulation map to another view, is useful during the exploration process for three reasons.

- It helps to isolate data to perform separate analyses.

- In the trajectory mode, the brush selects entire trajectories. When picking these trajectories, a new accumulation map is computed and unveils new accumulation initially hidden by the picked trajectories (image *d* in Figure 3.6).

- With the point and the trajectory mode, FromDaDy uses the full gradient scale in such a way that the minimum accumulation value has the first gradient color and the maximum accumulation value has the last gradient color. When brushing/picking and dropping points with minimum or maximum accumulation value, FromDaDy computes a new accumulation map that unveils a new maximum value with the maximum gradient color and then unveils new patterns (comparison of Figure 3.5 and Figure 3.6).

3.2.3 BRUSHING TECHNIQUE WITH 3D VOLUMES

Colortunneling (Hurter et al., 2014c) shows the most advanced technique to perform interactive brushing with large datasets. It uses advanced shading techniques with the render feedback buffer (Shreiner and Group, 2009) which will be detailed in Chapter 5. Color tunneling view layout is composed of two exploration views which show different visual mappings of the same dataset and a lock view (Figure 3.7).

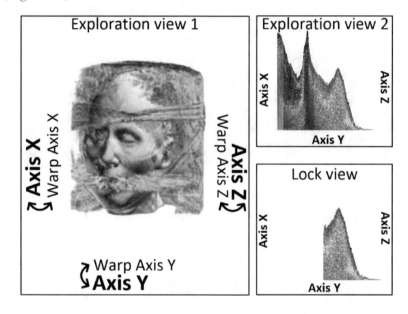

Figure 3.7: Color tunneling view layout (Hurter et al., 2014c).

While standard brushing techniques select or remove every brushed items (Figure 3.8), color tunneling used the lock view to only select or remove items which are not visible (i.e., locked) in the lock view. For instance, if the lock view shows a histogram of the density of a medical scan with only the high density (Figure 3.7), then the brushing technique will remove the low density (Figure 3.9) and show the human skull (high density).

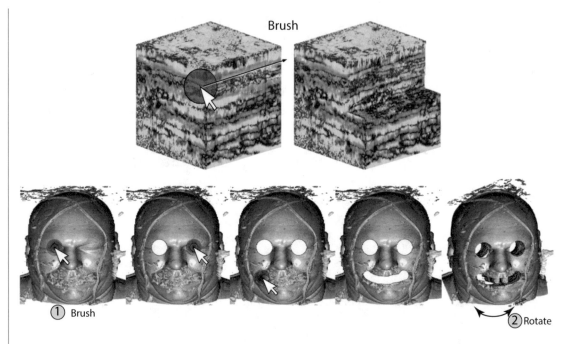

Figure 3.8: Standard brushing technique with a 3D datacube (top) and 3D medical scan (bottom).

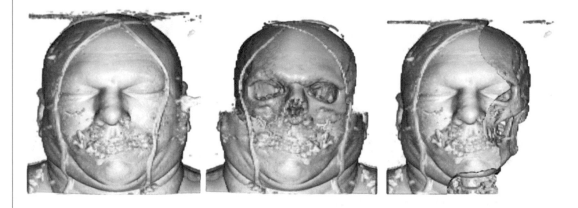

Figure 3.9: 3D medical scan with an advanced brushing technique that only removes low density voxels.

3.2.4   INTERACTIVE LIGHTING DIRECTION

In order to compute the shaded density map, one can consider it as a height map and use the standard Phong light computation (Phong, 1975). Since this technique needs a normal vector, a normal map can be generated thanks to the computation of the gradient of the density map (Figure 3.1).

The user can choose to display the accumulator map with, or without, this shading and interactively set the lighting direction with the mouse pointer. High accumulation values are considered as mountains that create shade, and low accumulation values are considered as valleys. By pointing with the mouse pointer to a specific area, the lighting direction can be interactively manipulated. This manipulation allows furrows or ridges to be emphasized. When defining the lighting direction from the left or right, vertical furrows are accentuated whereas a lighting direction from the top or bottom emphasizes horizontal furrows (Figure 3.10).

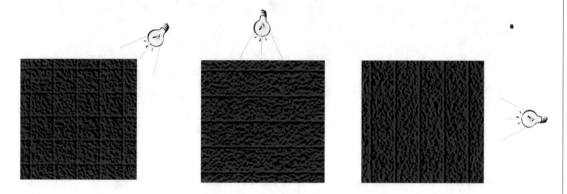

Figure 3.10: Interactive light manipulation to emphasise ridges and furrows on a shaded density map.

3.2.5   DENSITY MAPS AS DATA SOURCES

The color blending process computes an implicit density map since it combines pixels with the following blending formula:

OutputPixel = SourcePixel × SourceBlendFactor + DestinationPixel × DestinationBlendFactor; with 4D color vector (r,g,b,a) and the × symbol denotes component-wise multiplication.

Even if this formula can be customized, the pixels produced do not always provide an efficient quantitative comparison of the accumulation value. In Figure 3.11, the visualized database is a one-day record of aircraft flight plans (the routes that aircraft are supposed to follow). The view shows a matrix of points with the aircraft departure airport on the X axis and the aircraft type (Boing 747, Airbus A380, etc.) on the Y axis. Since many aircraft have the same departure airport and aircraft type couple, many points in the matrix have the same location on the screen. In the standard blended view, the brighter points show the most frequent couple in one day's traffic (Figure 3.11, left).

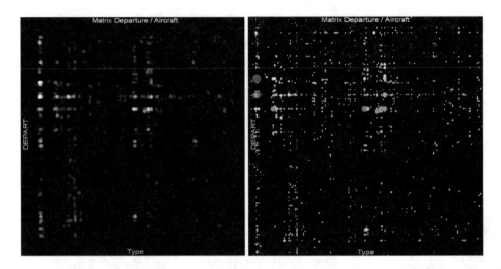

Figure 3.11: **Matrix view with standard color blending (left) and customized visual mapping with the size (right).**

FromDaDy offers another solution with a specific visual mapping. First, the user defines the data fields he or she wants to investigate (departure, aircraft type, etc.). Second, the system computes the corresponding density map. Finally, the user defines the visual mapping of the output view. In Figure 3.11, the density is mapped to the size and color. Figure 3.12 summarizes this configuration. This process operates as if a new field was provided into the dataset. The computed density map acts as a new data source.

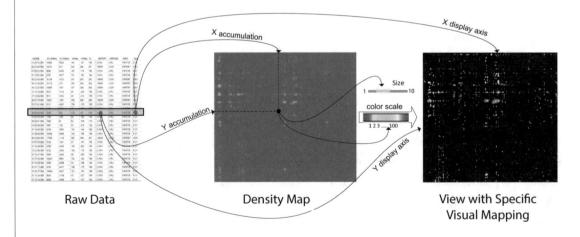

Figure 3.12: KDE maps value displayed with the size visual variable to display a matrix.

3.3   APPLICATION DOMAINS

This section gives some examples of usage in specific application domains. Additional examples are provided in a survey dedicated to ATC (Hurter et al., 2014b).

The first two examples show how density maps can make visual patterns appear. This can be useful to detect outliers or errors in a dataset. The two following examples are application domain related with the exploration of aircraft and gaze trajectories.

3.3.1   PATTERN DETECTION

This first example shows how density maps and their height map visualization can be used to isolate relevant aircraft trajectories. With the recording of one day's flights over France, each data represents the position of an aircraft at a certain time. The corresponding density map (Figure 3.13) is the result of the accumulation of plots with a triangular kernel. Hence, the X and Y position of each plot is mapped on the X and Y dimensions of the density map and on the X and Y dimensions of the resulting image. The image produced shows very dense areas over the main airports in France (Roissy, Orly, Lyon, etc.), which was expected.

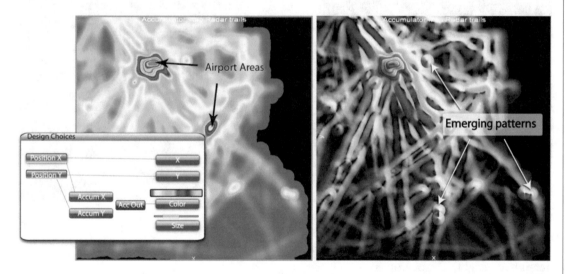

Figure 3.13: Design configuration and accumulation maps without shading.

When visualizing the density map with illumination, circular shapes emerge (Figure 3.13, right) that were not initially noticeable (Figure 3.13, left). Thanks to the shading process, density gradients are emphasized and this is the reason why these circular hills stand out. The user can then brush these shapes to extract the aircraft that causes this accumulation of data recording. Thus, the user brushes the hills and drops these data onto a second view. The user discovers that the picked trajectories correspond to stationary radar test plots recorded throughout the whole day. These radar test plots are mandatory to assess the correctness of the whole radar data processing (merging of multiple radar sources) (Renso et al., 2013).

In this second example, we use the data density as a tool to highlight flaws in the dataset. The dataset is a one-day record of aircraft positions. Radars send data over networks with a constant stream rate (in our dataset, one radar position of each aircraft every 4–8 min). Figure 3.14 shows the content of our dataset. The X screen axis shows the time of each radar plot and the Y screen axis shows the aircraft's identifier. Since the identifier of each aircraft is a number incremented over the day, the resulting shape shows a remarkable continuous pattern in which each horizontal line represents the lifetime of one flight (each flight has a unique identifier). The longest lines at the bottom of the visualization are the stationary radar test points recorded all day long. The width of this shape gives the average flight duration in the dataset: it is about 2½ hours which represents the average time to cross France by airplane (most aircraft cross France at a high altitude).

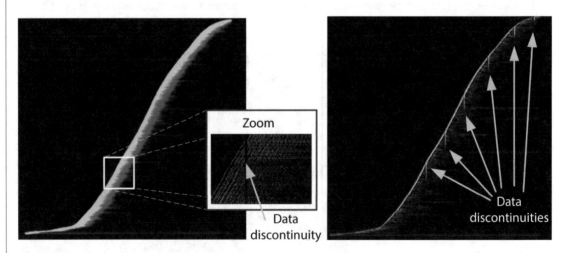

Figure 3.14: Time series of the incremental number of aircraft over time (left) and the corresponding density map (right).

Due to the large number of records, a lot of cluttering occurs when displaying the time series of the aircraft's identifier (Figure 3.14). This clutter hides the areas where, during a small time period, no data are recorded. Hence, the user is unable to discover this flaw in the database, unless by chance to zoom over the specific areas to reduce the cluttering of points, which is unlikely (Figure 3.14, zoom part). To notice this flaw without serendipity, the user can display the corresponding data density with shading (Figure 3.14, right). Thus, the density view unveils continuous and discontinuous data streams. A continuous data stream over time produces flat accumulations (the same amount of data are accumulated over time), whereas a discontinuous data stream produces ridges (increase of the data stream rate) or furrows (decrease of the data stream rate). If no data are recorded during a specific time-span, the produced accumulation view displays many furrows. Each of these furrows indicate that during the time corresponding to the thickness of the furrows, no data were recorded which reflects a failure in the recording system.

3.3.2   EXPLORATION OF AIRCRAFT PROXIMITY

The main activity of ATCs consists of maintaining safe distances between aircraft by giving clearances to pilots (heading, speed, or altitude orders). However, when aircraft fly below the safety distance, an alarm is triggered. These alarms are common since ATCs supervise aircraft in dense areas. Nevertheless, they are all monitored to avoid aircraft collision. In this example, the dataset contains only safety distance alarms with the recorded aircraft positions. The user connects the X and Y position of each aircraft to the X and Y density entries. The computed accumulation is visualized with a blue (low accumulation values) to red (high accumulation values) color scale.

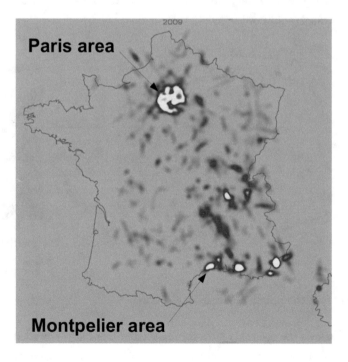

Figure 3.15: Density map of the safety-distance alarms triggered over France over a one-year period. Red-colored areas correspond to dense alarm areas where aircraft triggered proximity alerts.

Paris is, of course, the main dense area with the largest proportion of alarms. However, when visualizing all these alarms over a one-year period, users discovered that some unexpected dense areas emerge (Figure 3.15). For example, Montpellier, which is a far smaller airport, shows a lot of alarms. The user can use the selective brushing to retrieve the exact number of alarms.

3.3.3   EXPLORATION OF GAZE RECORDING

In this last example, I conducted an experiment that used an eye tracker. During the experiment, users were required to look at the center of the screen, then at a target located elsewhere on the screen, and finally to look back at the center of the screen. The database contains 200 eye trails with a total of 100,000 points. The accumulator map is used to reveal the speed of the user's gaze and eye fixation locations. To do this, the following configuration is used to compute the density map: trajectory Id is mapped to the Y axis, and curvilinear distance to the X axis (Figure 3.17a). The curvilinear distance corresponds to 0 at the beginning of the trails and increases until the trajectory ends. This distance is correlated with time, since trails are regularly sampled: hence, accumulation occurs between sampled points when the gaze speed slows down. The last step of the design is to retrieve its corresponding density for each gaze point. To do so, we map the density result to the size and color of the displayed lines (Figure 3.16).

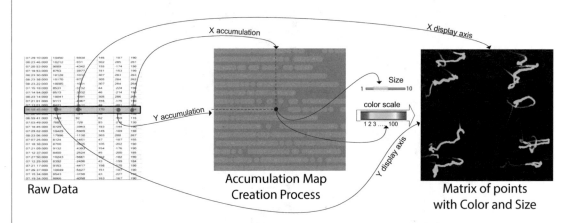

Figure 3.16: Density map computation with gaze recording.

Figure 3.17d shows the eye trails with standard color blending (recorded gazes are displayed with transparent dots): main stops are visible. Figure 3.17c shows the same trails with the use of the computed density map. The stops are visible with finer details thanks to the color scale and the variation of size. Thanks to the interactive brushing technique, the user can investigate specific trails in more detail.

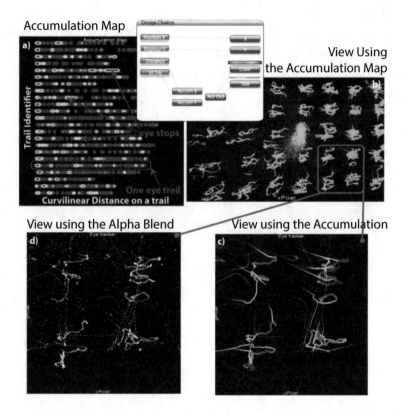

Figure 3.17: Accumulation view and its configuration to produce a per trajectory distance accumulation (a, b). Comparison between trail visualization with or without the accumulation (c, d).

3.4   CONCLUSION

This chapter showed how a simple image-based visualization with the density map computation can leverage data exploration. This image-based visualization is efficient only if interaction is also provided. Thanks to the computation power of the graphic card the following duality is possible: visualization and fast computation to support interaction.

The visualization is supported by a density map (i.e., raster map) which contains data densities and then at the very end of the transformation pipeline ("view") presents the data with different designs: 2D view with a color gradient, 3D view with a shading, etc.

The interaction is also provided thanks to raster maps but at a different stage of the visualization pipe line. The raster map can be stored in the "data table" (brushing on points) or in the "data structure" (brushing on trajectories). Other raster maps can be computed to dynamically update the shading when the user changes its configuration or when the user spread data after the brushing (pick and spread technique).

KDE map computation is one of the image-based visualizations. The next chapter will present others visualizations which are directly connected to the previous one with view simplification: edge bundling. The chapters that follow will also present additional image-based visualization with animations and distortions.

CHAPTER 4

Edge Bundling

In this chapter, we will discuss algorithms and interactive techniques to reduce graph/trajectory edge clutter in visualizations. These techniques mainly rely on edge simplification algorithms which have been subject to increased research interest and a number of improvements and enhancements in recent years. Edge bundling techniques can also be divided into the three levels from the data transformation pipeline (Figure 1.2): data level, geometry level, and image level.

Data Level Methods. On the data level, edges can be filtered or aggregated. Filtering techniques remove edges with given criteria, whereas aggregation techniques merge edges having similar semantics. Ellis and Dix (2007) give an aggregation and edge clustering methods. Interactive systems, such as Node Trix, compact dense subgraphs into matrix representations (Hadlak et al., 2011; Henry et al., 2007). In Ploceus (Liu, 2012), one can display networks from different perspectives, at different levels of abstraction, and with different edge semantics. Regarding filtering techniques, direct queries can filter out edges with multivariate criteria. Edges can be filtered with Centrality Based Visualization of Small World Graphs (van Ham and Wattenberg, 2008) or explored with spanning tree as used in TreePlus (Lee et al., 2006a).

Geometry Level Methods. On the geometry level, dense edge visualizations can be uncluttered by using edge bundling techniques. Edge bundling techniques trade clutter for overdraw by routing geometrically and semantically related edges along similar paths (Hurter et al., 2013b). This improves readability in terms of finding groups of nodes related to each other by tracing groups of edges (the bundles) which are separated by whitespace (Gansner et al., 2011). Dickerson et al. merge edges by reducing non-planar graphs to planar ones (Dickerson et al., 2003). The first edge bundling technique was the flow map visualizations which produce a binary clustering of nodes in a directed graph representing flows to route curved edges along (Phan et al., 2005). Flow maps' control meshes are used by several authors to route curved edges, e.g., Qu et al., (2007) and Zhou et al. (2008). These techniques were later generalized into edge bundling approaches that use a graph structure to route curved edges. Holten pioneered edge bundling for compound graphs by routing edges along the hierarchy layout using B-splines (Holten, 2006). Gansner and Koren (2007) bundle edges in a circular node layout similar to Holten (2006) by area optimization metrics. Control meshes can also be used for edge clustering in graphs (Qu et al., 2007; Zhou et al., 2008); a Delaunay-based extension called Geometric-Based Edge Bundling (GBEB) (Cui et al., 2008), and "Winding Roads" (WR) that use Voronoï diagrams for 2D and 3D layouts (Lambert et al., 2010b, 2010c). The most popular technique is the force-directed edge layout technique which use curved edges to minimize crossings and implicitly creates bundle-like shapes (Dwyer et al., 2007).

Force-Directed Edge Bundling (FDEB) creates bundles by attracting control points on edges close to each other (Holten and van Wijk, 2009) and was adapted to separate bundles running in opposite directions (Selassie et al., 2011). The MINGLE method uses multilevel clustering to significantly accelerate the bundling process (Gansner et al., 2011). Computation times for larger graphs struggle with the algorithmic complexity of the edge bundling problem. This makes scalability the major issue when using edge bundling techniques. Therefore, the latest techniques use therefore the image level to bundle edges.

Image Level Methods. Thanks to the recent improvements regarding graphic hardware and its flexible usage, image level methods are very popular nowadays. Graphic cards can be used to improve rendering aesthetics and address scalability issues. Several techniques exist for rendering and exploring bundled layouts: edge color interpolation for edge directions (Cui et al., 2008; Holten, 2006); and transparency or hue for local edge density, i.e., the importance of a bundle, or for edge lengths (Lambert et al., 2010a). Bundles can be drawn as compact shapes whose structure is emphasized by shaded cushions (Scheepens et al., 2011; Telea and Ersoy, 2010). Graph splatting visualizes node-link diagrams as continuous scalar fields using color and/or height maps (Van Liere and Leeuw, 2003; Hurter et al., 2009b). Several techniques exist to improve scalability based on image level. Skeleton-based edge bundling (SBEB) uses the skeletons or medial axes of the graph drawing's threshold distance transform as bundling cues to produce strongly ramified bundles (Ersoy et al., 2011). To explore crowded areas where several bundles overlap, bundled layouts can be interactively deformed using semantic lenses (Hurter et al., 2011). Hurter et al. (2013b)use a pixel based bundling method to explore dynamic graphs.

All of the image level methods are image-based visualization. Methods that especially rely on edge bundling will be chronologically presented in the following. Since they apply image-based techniques they are scalable with a low complexity and have already proven to reduce clutter in dense graph and trajectory visualization (Hurter et al., 2014b).

4.1   SBEB: SKELETON-BASED EDGE BUNDLING

One of the very first image-based bundling techniques was a follow-up of a previous work on edge visual improvements (Telea and Ersoy, 2010). Prof. Telea and Wijk (2002) intensively investigated skeleton computation thanks to different techniques, mainly distance transform and feature detection. Thanks to this skeleton computation, it is possible to compute the distance of each edge from the center line. One can then apply a gradient color to emphasize the borders of the bundles and thus greatly improve the visualization of bundles (Figure 4.1).

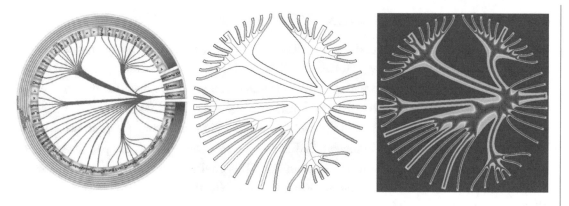

Figure 4.1: Left: edge layout; center: the corresponding skeleton; right: gradient profile.

This image-based technique has been extended to support edge bundling computation. Rather than using this center line having a distance computation, it can also be used as a magnetic guide to attract edges (Ersoy et al., 2011). Even if this technique remains simple, additional computations have to be performed to ensure high-quality bundles (kinks removal and relaxation, i.e., slightly unbundle the graph).

This bundling technique is a pixel-based technique which computes the skeleton of a splatted graph and attracts edges toward it (Figure 4.2). A clustering stage is mandatory to produce detailed bundles. This algorithm only needs two parameters to make it work: an attraction and a smoothing factor. Some post processing can be performed to improve the visualization such as edge borders (Telea and Ersoy, 2010).

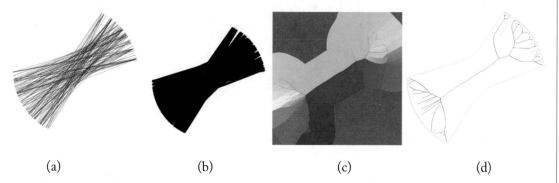

| (a) | (b) | (c) | (d) |

Figure 4.2: Skeleton computation. A set of edges (a) is inflated (b), then the distance tansform is computed (c), and finally the skeleton is extracted (d).

4.2   KDEEB: KERNEL DENSITY EDGE BUNDLING

The main concern with edge bundling algorithms was their complexity and their computation time. Hierachical Edge Bundling (Holten, 2006) is the fastest algorithm but needs a data hierarchy which drastically restricts its usage. FDEB (Holten and van Wijk, 2009) is flexible enough but its computation time is prohibitive when using a large dataset. SBEB (Ersoy et al., 2011) is scalable, but needs data clustering and the skeleton computation takes time (one can use CUDA, but its implementation remains complex). SBEB was the first algorithm to use pixel-based algorithm: the skeleton and its gradient map computation. This gradient map will attract edges toward the medium skeleton axis. Kernel Density Edge Bundling (KDEB) idea is simple (Hurter et al., 2012): since the skeleton and the gradient were the complex and the time-consuming parts of this algorithm, one needs to find another solution to compute them. The density map and its corresponding gradient are very good substitutes to the skeleton map computation. When attracting an edge toward a dense area, it would make them overlap and thus create a denser area but also empty spaces. This iterative algorithm may produce a clearer view and a new bundling algorithm. Open question remains to assess if this algorithm can be stable enough to converge and avoid too strong distortions. The first prototypes produced very bad bundling results, but in practice this algorithm converged (Figure 4.3).

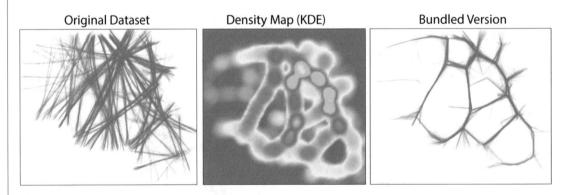

Figure 4.3: First bundling prototype based on density computation.

This algorithm ends up producing high-quality bundles thanks to many improvements: adding an edge resampling algorithm, an iterative density map computation, and a smooth edge advection. Since this algorithm uses a density map, we can directly use this edge overlapping information (i.e., density) to improve the visualization with, for instance, bump mapping (Figure 4.4). This algorithm was first developed without formal explanation to validate how it converges. One year after its publication, Hurter et al., (2013c) found out that the mean shift algorithm had already explained its convergence in a different application domain: data clustering (Comaniciu and Meer, 2002).

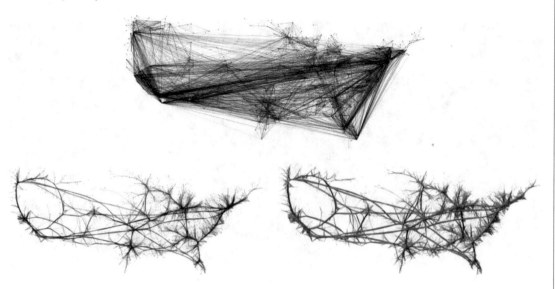

Figure 4.4: U.S. migration graph. Original (top), bundled (bottom), with shaded density (bottom right).

4.3 DYNAMIC KDEEB

Many additional investigations took place after the first publication of KDEEB (Hurter et al., 2012). The purpose of one of them was to discover why this algorithm converges (Comaniciu and Meer, 2002), and why it is noise resistant: with the same parameters (kernel size, attraction, interaction, smoothing, resampling) a bundled graph remains almost the same even if a small number of edges are added or removed. This property is a great asset, especially for producing a bundled version of a dynamic graph. A dynamic graph has a given amount of edges that change over time (removed, added, or displaced). Previous attempts only managed to produce dynamic graph visualization with large jumps between key frames, and with reduced datasets (Nguyen et al., 2013a). KDEEB uses an accumulation map which will not drastically change over time and thus ensures continuity of the bundled dynamic graph. Furthermore, KDEEB uses a pixel-based technique which is highly scalable, and is probably the fastest bundling algorithm (except for Hierarchical Edge Bundling (HEB) (Holten, 2006) which only displays splines without additional data processing). With the algorithm assets, nothing prevented us anymore from producing the first bundling algorithm of a dynamic graph. This initial publication depicts the dynamic bundling principle with airlines and software visualization dynamic graph exploration (Hurter et al., 2013c). And thanks to a new interest in eye tracking systems, Hurter et al. (2013a) also discovered a very promising usage of bundling techniques with eye gaze. As an example, one can bundle the pilot's gaze during a landing sequence and discover the standard monitoring pattern (Figure 4.5).

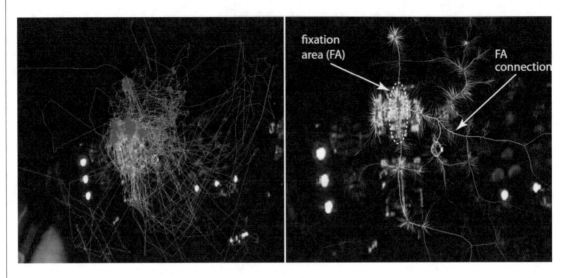

Figure 4.5: Eye gaze recording of a pilot when landing with a flight simulator. Bundled trails (right) with KDEEB.

4.4   3D DKEEB

There is no restriction to extend the KDEB to take into account 3D graphs (x,y,z). This work is a simple extension of the original algorithm (Hurter et al., 2012) in which a 3D density map is computed with a 3D gradient. The computation time drastically increases due to the additional volume of data. The density map becomes a 3D texture (e.g., 400^3 voxels) and the gradient becomes a 3D vector.

The results are visually compelling by reducing the clutter of the 3D visualization of aircraft trajectories (Figure 4.6). Space-time cube bundling is also not subject to any technical limitation: the z axis can directly be mapped to the time dimension.

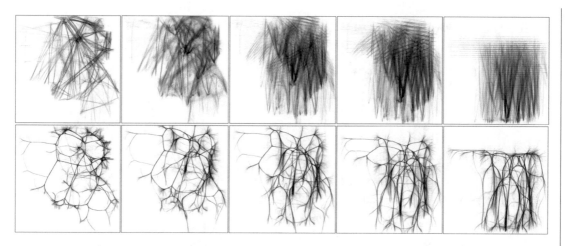

Figure 4.6: 3D DKEEB with one day of recorded aircraft trajectories. The top left image shows the geographical visualization of the dataset. The top right view shows the vertical visualization of the traffic (the altitude is mapped on the y axis). The bottom images show the corresponding 3D bundled version of the dataset.

4.5   DIRECTIONAL KDEEB

KDEEB (Hurter et al., 2012) has proved to be a very promising algorithm as it is scalable and convergent. However, one limitation remains in the usage of direction graph/trajectories. When visualizing an aggregated view of aircraft trajectories, it is nonsense to bundle opposite direction trajectories. To solve this issue, one can use preprocessing clustering. Peysakhovich et al. (2015) suggested a simple extension of KDEEB to take into account the direction. Instead of using a gradient map, one also computed an accumulation direction map. For each pixel in a raster map one computes the average direction of the edge which would be located over this pixel. To apply the gradient, one first tests the compatibility of the edge direction and the average direction. If they were close, the gradient is applied; if not, the edge does not move. This extension does not hinder the algorithm complexity which remained linear O(E). In addition, this algorithm manages to take into account the edge direction. This algorithm is GPU implemented to help scalability. Comparison of directional and standard bundling technique is give in Figure 4.7.

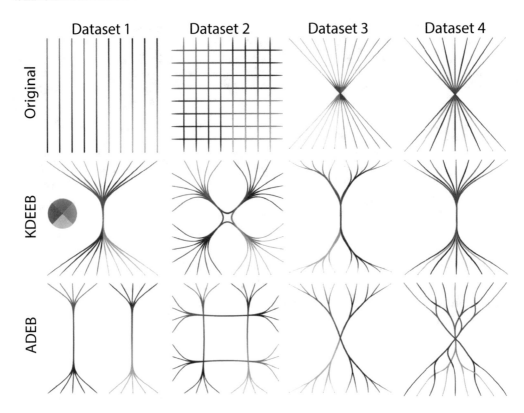

Figure 4.7: Comparison of KDEEB and ADEB technique when dealing with oriented graphs. The color of the edges corresponds to their direction.

Figure 4.8 shows the global data visualization of aircraft trajectories with the Direction-KDEEB technique. Trails with the same direction have been bundled. The color corresponds to the global trail direction; ADEB computed the direction for each trail with its start and end point. This is valid since during a flight an aircraft can change direction (follow a specific flight route, avoid dense area of traffic, apply ATC orders), but each aircraft has a main direction given by the start and end points. Figure 4.8 shows a color coding with a bright color at the start and a dark one at the end point; in this way we emphasized the visualization of outcoming and incoming bundles. As such, the London and Paris areas contain numerous incoming and outgoing flows, which makes these areas dense and complex.

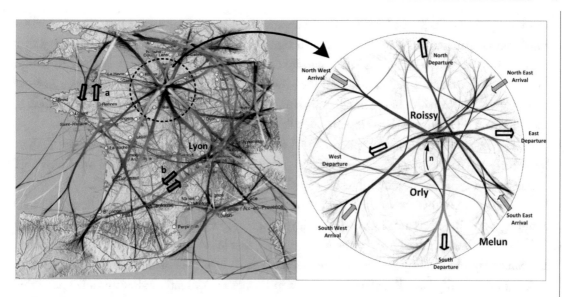

Figure 4.8: Investigation of aircraft trails with a directional bundling algorithm (Peysakhovichet al., 2015).

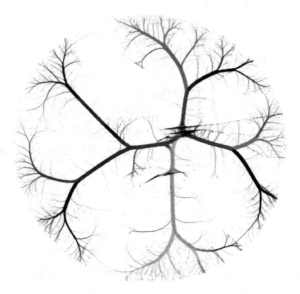

Figure 4.9: Paris area with the KDEEB algorithm (Hurter et al., 2012).

The visualization also highlights the complex configuration over the Lyon Area (see Figure 4.8, left). The Lyon area is a central crossing for European flight routes which does not have the same flow configuration as Paris. Whereas the Paris area does not have transit aircraft (aircraft that fly over a given area without landing), the Lyon area does have major transit flows (e.g., Switzerland/France/Spain, arrows b in Figure 4.8). In terms of traffic management, one can observe that many airways are grouped in pairs with opposite directions. This is especially the case with arrows to reach and to exit the London area, and the airways between Switzerland/France/Spain (arrows b in Figure 4.8). The width of the bundles also highlights traffic density between aircraft. Arrows b, trails between Switzerland/France/Spain, are balanced in their two directions.

In Figure 4.8 (right), we applied ADEB algorithm on a subset of the trail dataset. Due to the density and dataset size of aircraft trajectory, no previous investigation had been previously able to bundle and extract a meaningful result, especially over the Paris area (Figure 4.9).

Figure 4.8 (right) shows the specific flow configuration with four incoming flows and four outgoing ones. The Paris area has two main airports: Roissy Charles de Gaulle (CDG) and Orly (ORY). This visualization shows, by the traffic density, that Westerly departures are the least dense, and Southerly and Easterly are the most dense. Since Paris is in the north of France, local traffic is mainly to the south. Easterly traffic corresponds to European destinations. Specific analysis of this visualization shows that Orly, the second biggest airport in France, has a strong departure flow to the south but a reduced one to the north (barely visible on the map with the arrow n). Since Orly is a domestic airport (flights remaining in France) located in in the north of France, only a few aircraft head to the north.

4.6   EDGE BUNDLING GENERALIZATION

Even if this directional bundling technique takes its root on top of the KDEEB algorithm, it remains a new way to bundle edges thanks to the image-based technique.

Edge bundling generalization introduces a new density map which contains a local edge measurement computation. The complexity to compute this map is $O(n)$ which ensures the scalability of this algorithm. Furthermore, this algorithm embeds a compatibility measurement to attract edges with close proximity and direction. This initial edge compatibility measurement can be easily extended to support any attribute compatibility. One can display the computed density and compatibility map (Figure 4.10).

To enable the edge bundling generalization, one has to map any attribute to an angle measurement. This simple mapping transforms $a \in [a_{min}; a_{max}]$ to a polar coordinate v_a with the following formula:

$$v_a = \begin{cases} x = \cos\left(\dfrac{a - a_{min}}{a_{max} - a_{min}}\right) \\ y = \sin\left(\dfrac{a - a_{min}}{a_{max} - a_{min}}\right) \end{cases}$$

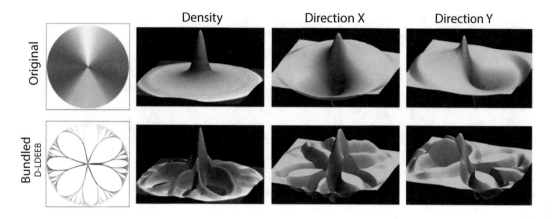

Figure 4.10: Original and bundled dataset with its corresponding density maps and orientation densities (direction X and Y).

This generalized edge bundling technique can be applied to many application domains like graph simplification, trail and gaze analysis, etc. (Peysakhovich et al., 2015).

4.7 DENSITY COMPATIBILITY

Since KDEEB and ADEB use a density map to apply edge aggregation, the edge density is directly computed and can be displayed to emphasize the visualization. Each edge can be then displayed with a width which corresponds to the local edge density. The general result is visually satisfactory (Figure 4.11) but many artifacts may appear (Figure 4.12, top left). These artifacts correspond to crossing edges with incompatible directions.

Thanks to the compatibility map provided by ADEB, one can only assign the width of an edge if its local density is compatible in terms of orientation (Figure 4.12, top right) (Peysakhovich et al., 2015).

Figure 4.11: Artifact when using the edge width with the density map (top left); artifact removal when taking into account the compatibility map (top right).

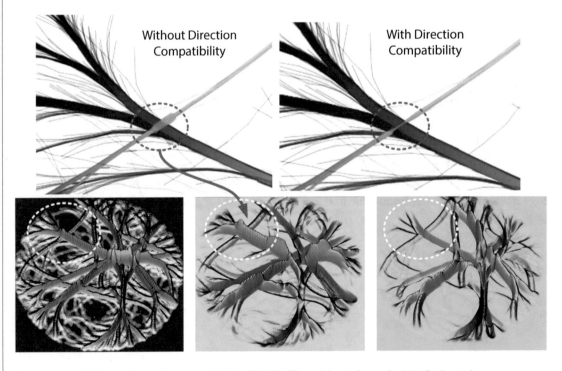

Figure 4.12: U.S. migration dataset with the ADEB (Peysakhovich et al., 2015) algorithm.

4.8   PROPOSAL TO IMPROVE BUNDLING TECHNIQUES

The following list shows potential improvement of existing edge bundling techniques.

- **Provide or evaluate metric bundle links:** Today, there is no metric to quantify the result of a bundling algorithm. The only metric available is the ration ink/background proposed by Tufte (1986), which is far from fully qualifying the result of an algorithm.

- **Provide reference data sets:** Each new implementation of a bundling algorithm uses new datasets which makes their comparison difficult. Mingle (Gansner et al., 2011) used sets of reference data (Davis and Hu, 2011) but provided no indication of any information to be extracted. It would therefore be useful to collect a set of validated data sets with ground truth data such as reference calculation speed, rendering quality, and data mining references.

- **Link bundling algorithms to tasks:** To improve edge bundling usage, one should be able to choose suitable techniques which match best the requested graph simplification tasks. This is a difficult problem since each algorithm has technical constraints that are not correlated to specific tasks. For example, hierarchical data is mandatory to apply HEB (Holten, 2006), data clustering must be applied before processing them with SBEB (Ersoy et al., 2011), and FDEB (Holten and van Wijk, 2009) needs rules of proximity.

- **Improve the parameterization of algorithms:** Each bundling algorithm has its own complexity with a different set of parameters. HEB (Holten, 2006) is by far the simplest, but also the most constrained with the need for a hierarchical data structure. Mingle is by far the most complex to implement with the use of the GPU to export a dependency graph. All these algorithms use many parameters for their computation (except HEB that uses only the B-Spline parameter). These parameters are often too abstract and related to the technical constraints of the algorithm. For example, KDEEB (Hurter et al., 2012) uses the size in pixels of a density map that has no direct connection with the exploration task of a graph. Finally, each bundling technique should be provided with a set of reference data sets and their set of optimal parameters. For example, KDEEB uses multiple bundling iterations with changing parameters.

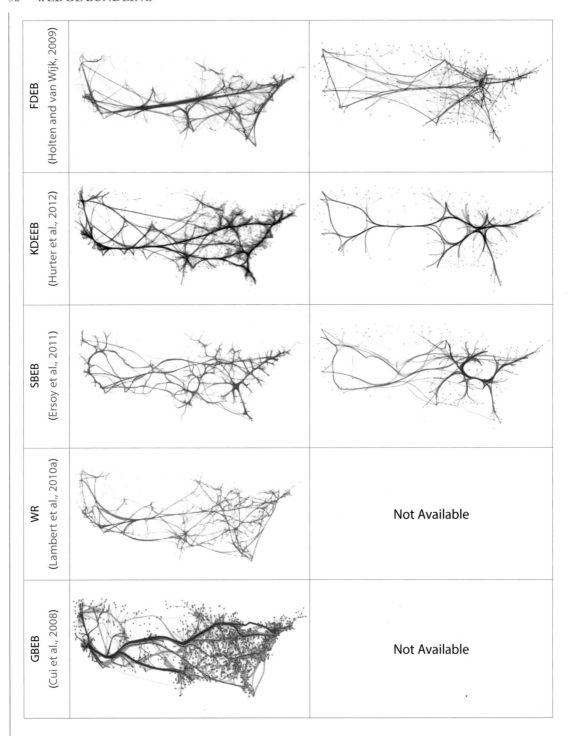

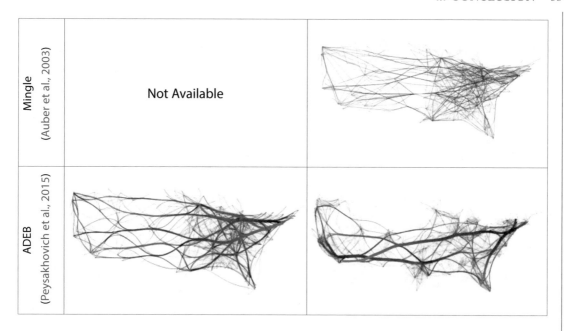

Figure 4.13: Examples of bundling results with very different visual renderings on the same datasets.

4.9   CONCLUSION

This chapter listed many graph/trajectory simplification techniques and focused on the image level methods. These image-based methods have proved to be scalable and produce valuable simplified visualizations. This work also contributes to the dense graph visualization and provided a definition for the edge bundling algorithm:

Edge bundling techniques trade clutter for overdraw by routing related edges along similar paths.

Graph visualization supports various comprehension tasks such as understanding connectivity patterns, finding frequently taken communication paths, and assessing the overall interaction structure in relational datasets (Lee et al., 2006b). Much further work is required to fully understand how edge bundling algorithms support such tasks. Current and future challenges and extensions of image-based edge bundling techniques will be discussed in the last chapter of this book.

In the next chapter, we will investigate another feature of image-based visualization with the animation between views. As a simple example, the mole view (Hurter et al., 2011) shows the animation between a graph and its bundled version thanks to a user-controlled lens.

CHAPTER   5

Animation with Large Datasets

This chapter investigates other uses of pixel-based visualization with animations. Animations have many assets (Tversky et al., 2002): they provide an aesthetic venue when exploring datasets and a great deal of previous research has shown that animations can help a user to investigate a dataset by providing non-disruptive artifacts when modifying view configuration (Bezerianos et al., 2010; Elmqvist et al., 2008; Hurter et al., 2009b). Previous works also mentioned that the user mental map is somehow preserved (Archambault et al., 2011) which greatly reduces the cognitive workload and helps him or her to focus on the goal of the data exploration: finding relevant information. In the same way, this chapter will investigate how animations and view distortions can help when investigating multidimensional datasets.

Both animation and distortion share the same challenge: How to make them interactive with large datasets? Standard visualization methods can display and animate few visual entities, but when dealing with more than 1,000 items, systems often show animation lock-ups and cannot animate the views with a suitable frame rate to support fluid transitions (around 12 images per second). Therefore, image-based visualization is a perfect candidate to improve data exploration (Card et al., 1991) and to address scalability issue.

This chapter will present a journey which progressively details more challenging animation issues. It will start with the animation between view configurations, bundled-unbundled layout, image and its histogram, and particle animation, and end with lens distortions in 2D and 3D.

5.1   ANIMATION BETWEEN DUAL FRAMES

When visually exploring a multidimensional dataset, users may need to use multiple view configurations in order to find information. For example, if one displays data with a scatterplot, users may change the mapping between the dimensions of the data and the X and Y coordinates of the scatterplot. When this change occurs, the system can switch views instantaneously or use an animated transition. When switching instantaneously, the user optical flow is broken, hindering its capability to relate the graphical marks used to plot the data between the initial and final views (Shanmugasundaram et al., 2007a; Heer and Robertson, 2007). This results in the inability to perceive and understand the relationships between the data. Conversely, the switch between the initial and final view can be done using a smooth animation that moves the graphical items from their position in the initial view to a new position in the final view. With such a technique, users may be able to track visual items and understand the relationships between them. Various animations have been proposed to help users figure out the switch between views (Heer and Robertson, 2007).

5.1.1   ROTATION TO SUPPORT DUAL SCATTERPLOT LAYOUT

One way to perform a smooth transition is to use a temporary 3D view that rotates like a transparent die (Bezerianos et al., 2010; Elmqvist et al., 2008; Hurter et al., 2009b). In the case of aircraft trajectory visualization, users would rotate the view from a 2D trajectory top view to a 2D vertical view around the X axis (Figure 5.1). The assumed benefit of the technique is that the rotation and accompanying changing visual structure are easier to understand since it is ecological (Fitzmaurice et al., 2008), i.e., humans, as living organisms, have evolved and tuned their perceptual systems to perceive occurring events in nature, such as rotations (a rolling boulder, a fruit or tool in a hand, etc.).

FromDaDy (Hurter et al., 2009) supports such animation: even if more than 400,000 connected points are animated the fixed pipeline of the graphic card can smoothly animate such a dataset. In order to improve such animation, it is preferable to let the user control the animation rather than an automatic rotation.

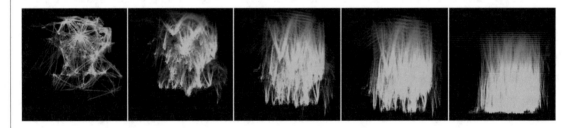

Figure 5.1: One day of aircraft trajectory with an animation between the top view (latitude, longitude) and the vertical view (altitude, longitude).

Previous work investigated such animation to better understand how the rotation axis helps in the data understanding (Cordeil et al., 2013).

5.1.2   ANIMATION BETWEEN AN IMAGE AND ITS HISTOGRAM

An image is composed of a set of juxtaposed colored pixels. A luminosity histogram is a visualization than represents a bar graph with each bin representing the number of pixels that share the same luminosity. These bins are ordered by increasing luminosity values. The very first animation between an image and its corresponding luminosity histogram was presented during the oral presentation of the MoleView (Hurter et al., 2011) at the conference InfoVis 2011 (Figure 5.2).

Figure 5.2: First prototype of an animation between an image and its corresponding histogram.

Figure 5.2 shows an animation in which every pixel is animated from its original location in the image toward the maximum value of its corresponding bin in the luminosity histogram. An improved animation has been implemented in Histomage (Chevalier et al., 2012), a novel image manipulation software. Histomage uses the fixed graphical pipeline to support this animation. Histomage can handle images up to a certain size, but images more than 2 mega pixels in size cannot be animated interactively. This limitation is linked to the intrinsic Histomage visualization process and suffers from the same limitation as the Mole View (Hurter et al., 2011). Since the animation is applied to every pixel of the image, the computation time is directly linked to their number.

A GPU version was also implemented thanks to the vertex shader which will compute the current position of every pixel. The interpolated position P_c is computed thanks to a linear interpolation between the location of the pixel in the image P_i and in the histogram P_h:

$$P_c(t) = t\,P_i + (1-t)\,P_h \text{ with } t \in [0,1]$$

Figure 5.3: GPU animation between an image and its corresponding luminosity histogram.

For histogram views, several extensions of the classical 2D histogram exist. A histogram view can be parameterized by up to four attributes (*v, c, h, d*) as follows:

- *v* is depicted along the histogram x axis. All points having $v = x$ are stacked along the y axis at *v*;

- *c* determines the color of each point;

- for all points with the same v value, h gives the sorting order along y; and

- d determines the (back to front) order in which points are drawn.

These mechanisms effectively help in finding patterns of interest in histogram views. Figure 5.4 shows a (purposefully simple) example, where the dataset F is a 2D color image. Figure 5.4b shows a luminance histogram, with same-luminance points sorted vertically on chrominance and same-luminance and same-chrominance points sorted, in depth, on hue. We observe three quite well separated, vertically stacked, color bands (blue, orange, and pale green). This view is thus useful for selecting image portions based on their chrominance. Figure 5.4c shows a luminance histogram, with same-luminance points sorted vertically on hue, and same-luminance and same-hue points sorted, in depth, also on hue. The orange cluster corresponding to the fish in Figure 5.4a appears as a well-isolated narrow band at the top of the plot, so this view can be used to select this image structure. Finally, Figure 5.4d shows a luminance histogram, with same-luminance points sorted vertically on their original y coordinate, and same-luminance and same-y points sorted, in depth, on chrominance.

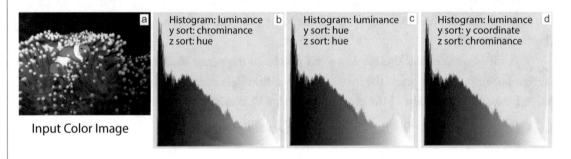

Figure 5.4: Color image (a) and three types of histograms (b).

5.1.3   INTERPOLATION BETWEEN TWO VIEWS WITH LARGE DATASET

Thanks to GPGPU techniques, the scalability issues when visualizing and interacting with large datasets can be addressed. In order to extend the Mole View and Histomage techniques (Chevalier et al., 2012; Hurter et al., 2011), the feedback render buffer can take advantage of one of the most advanced GPU usages. This new type of shader technique (programmable graphic pipeline) allows one to store in GPU memories the output of the geometry shader. Usually, the programmable pipeline modifies the geometry of vertexes and uses the raster map to display it. This computation needs to be performed at every frame. The rendered back buffer allows the storing of vertex modifications and thus optimizes the computation of vertexes geometry. This process is mainly used to compute the location of particles in particles systems, but other usages are possible. Since the Mole

View uses point sprites (close to particle system), this process managed to speed up the geometry computation between two consecutive frames. Such a prototype can animate up to 20 mega pixels with a frame rate of more than 20 fps on a 480 GTX nvidia card. This technological improvement was the missing factor to address the scalability issue with the color tunneling (Hurter et al., 2014c). Color tunneling can handle the animation between one image and its corresponding histogram or other pixel structure (Figure 5.7) up to 15 million pixels.

Color tunneling has proven to be efficient enough thanks to specific interaction features. First the user can animate the transition between two visual configurations. Second, the user can control the animation and stop at any stage to better track an object or detect visual patterns. Third, the user can brush to select or remove visual entities in order to remove or reduce occlusion. The following examples show concrete usages of such features.

Consider the 3D scans in Figure 5.5 and Figure 5.6 (128x128x112 voxels). We want to "peek" inside the head to see the brain. The animation between the 3D scan and the gradient view show the brain and help to track it but not to select it (Figure 5.5).

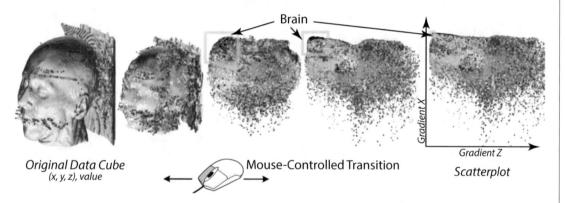

Figure 5.5: Animation of a medical scan (3D) to a scatterplot (gradient view). This animation helps to track the brain location between the two views.

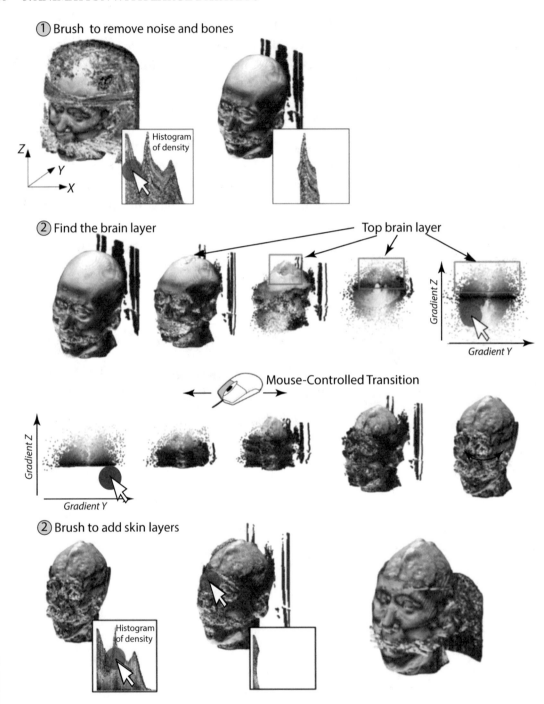

Figure 5.6: Exposing the top part of the brain structure in a 3D scan.

Figure 5.6 shows the brain extraction scenario. Here, we want to expose the top part of the brain structure in our head scan. Simple filtering cannot easily achieve this. The human head in this scan consists of a succession of layers with non-monotonic density values (low for skin, high for bone, and low again for the deeply nested brain structure). Simply filtering out the bones will not help to display the brain since the skin will not be filtered and will occlude it. Conversely, filtering on the skin density will also remove the brain which has a similar density value. To solve our task, we use color tunneling. First, we use a density histogram view and erase the noise (low density) and bone (highest density) values (Figure 5.6-1). Next, we create a 2D scatterplot of the z value of the density gradient vs the x gradient value, and use the warp tool to animate between the 3D DVR and this scatterplot (Figure 5.6-2). Warping a few times back and forth, we see that the top part of the brain is warped to the top half of the 2D scatterplot. This matches the fact that, in this area, z density gradients are large. We now remove the lower part of the brain by erasing the scatterplot's lower half (Figure 5.6-3, right image). However, this also erases some skin parts. To get these back, we use the density histogram view to unlock points in the skin density range (Figure 5.6-3, left). Finally, we use the add brush in the DVR view to paint back the skin voxels in the damaged areas (Figure 5.6-3, middle). Since only soft-density voxels are unlocked for editing, and we brush only over skin areas, only skin voxels get affected; bone or noise voxels are not painted back. Figure 5.6-3 (right) shows the final result.

5.1.4   THE ANIMATION AS A TOOL TO DETECT OUTLIERS

As explained in the previous section, the animation can help to explore multidimensional datasets. In this section, we will present a specific scenario in which animation will emphasis outliers.

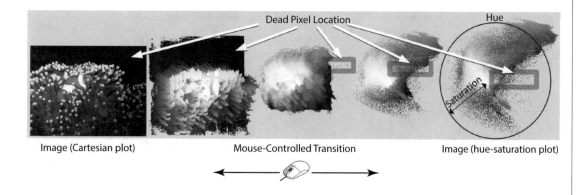

Figure 5.7: Color tunneling animation between an image and its corresponding histogram (top) or another color layout (bottom) (saturation, hue).

Consider a 2D color photograph, shown as a Cartesian plot (Figure 5.7 left). Photos often contain isolated pixel groups whose color slightly differs from their surroundings, such as "dead pixels" due to imperfections of digital cameras. Isolating such pixels (e.g., for retouching) is hard: they are visible neither in Cartesian nor in hue-saturation plots (Figure 5.7 right). However, if we warp between the two plots such pixels clearly show up as outliers (Figure 5.7, middle frame). Why does our animation highlight such outliers? The explanation is as follows. Similar-colored compact spatial regions in the Cartesian plot, e.g., the uniform image background or the orange fish, move as compact blocks to their corresponding regions in the hue-saturation plot. Outlier pixels in such regions have different hue and/or saturation values, so they are warped on different trajectories. In our case, these are pixels on the dark image background, whose color slightly differs from their uniform vicinity. We can stop the animation at any frame showing such pixels to select them for, e.g., retouching.

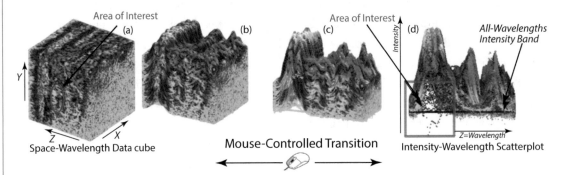

Figure 5.8: Color tunneling (Hurter et al., 2014c); finding intensity outliers with isolated ranges in an astronomical data cube (Taylor et al., 2003).

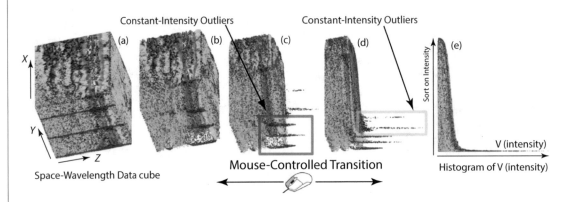

Figure 5.9: Locating constant-intensity line outliers in an astronomical data cube (Hurter et al., 2014c).

Next we consider a 3D cube of astrophysical measurements of the largescale structure of hydrogen gas intensities in our Milky Way Galaxy (1024x1024x160 16-bit integer voxels, 320 MB total) (Taylor et al., 2003). The x and y axes map polar sky coordinates, and z maps radiation wave-length, which translates to distance through the Galaxy along ray paths. Color maps gas intensity. Figure 5.8a shows our data cube, rendered with DVR. In this view, astrophysicists using our tool, and who provided the feedback outlined in this section, could only see color layers that indicate regions of denser hydrogen gas from the spiral arms of the Milky Way, such as the prominent yellow slab spread over a large part of the xy subspace. These are regions of the Galaxy where the cycles of star birth and death play out. We now choose a scatterplot of intensity vs. wavelength (Figure 5.8d). This shows two interesting phenomena. First, we see a thin compact horizontal black bar, not visible in the initial data cube. This tells that the respective intensity is present in all wavelengths. Second, we notice a white gap in the intensity-wavelength space, above the black bar at the distance of the bright spiral arm (Figure 5.8d, red marker). This tells that, for the respective wavelengths, there exist only high (purple:yellow) intensities, but no intermediate (blue:green) intensities. This situation does not occur for any other wavelengths, as there is a single such gap in the scatterplot. We now warp the scatterplot towards the original data cube: the intermediate frames (Figure 5.8 b, c) show that the gap corresponds to the spatial region marked in red in Figure 5.8a, right inside the yellow wavelength band. This lack of low intensities at the location of the spiral arm shows an absence of low-density hydrogen in this region. Some phenomena may have swept up the gas into high-density structure, a step on the way to forming new stars.

Figure 5.9 shows a second scenario. In the DVR image, we notice a few constant-intensity lines parallel with the wavelength (z) axis (Figure 5.9a). Such lines are created by radiation from objects in the far universe being absorbed by gas in our Milky Way. The properties of these lines can be used to measure the galaxy temperature. We would like to select such lines for closer analysis. Doing this via spatial or value-range filtering is hard, since the lines are embedded in surrounding data, and also do not have a perfectly constant intensity. Also, we would like to find if similar lines exist deeper in the data cube. We can select these lines as follows. First, we build a histogram of the intensity gradients of our data points. Gradient is a good detector for the boundaries of these lines, as intensity rapidly changes between the relatively constant value inside lines and varying values out-side. We next sort histogram points vertically based on their intensity value, and order them in depth with high-intensity voxels first. (Figure 5.9e shows the result.) We see that relatively few voxels have high gradients, while the vast majority of the data is represented by a well-defined distribution of gradients. Our lines of interest are located in the former voxels (histogram tail). Also, we see several color bands in the smooth part of the histogram, with a thin purple (high-intensity value) band at the top, and most points having low values (green). Such bands emerge because of our y sorting on intensity. This distribution reveals the kinematics of the galaxy and the velocity structure of the gas, represented by gradients in intensity with wavelength. Over the high-gradient tail, we mainly see the same green shade as on the lines in Figure 5.9a. This indicates that, for these high gradients, points do not have high intensity values (purple). Our lines of interest thus occupy regions of low-intensity values and high gradient. To find our desired lines, we now warp between the histogram and DVR views. In the intermediate frames (Figure 5.9b-d), we see several horizontal lines appearing, which smoothly move from the histogram tail towards their spatial locations in the DVR view. To select all such lines, we thus simply select the histogram tail. For more control, we can use the transition views to select any desired line located at specific spatial positions. The animation unearths several additional such lines inside the data cube, which the DVR view (Figure 5.9a) did not show.

5.2   ANIMATED PARTICLES

In this section, we will investigate another animation asset when exploring large dataset with animated particles. Since trajectories of moving objects are directed, the direction information is highly relevant and important to display. Most existing systems use arrows to show direction (Andrienko and Andrienko, 2008), leading to cluttered views that hinder data exploration. Other solutions investigated the use of color gradients; this requires specific visual mappings, and subsequent color blending gives issues when gradients overlap. Some systems employ animated textures to show direction (Blaas et al., 2009), requiring a minimum trajectory width, which is not suitable for large data sets with many entangled trajectories.

One solution to show edge direction can use a color coding (Figure 4.8) but other solutions exist thanks to animated particles. Particle systems are an under-exploited visualization technique that show the spatial extent and direction of a trajectory through particles and can provide strong cues with little clutter. Even if trajectories overlap, particle movements remain visible and indicate trajectory directions. Moreover, combined with particle density indicating trajectory densities, traffic flows become visible. As a drawback, a particle system requires animation with a high frame rate and interactive response time. The latter requirements are especially challenging for large moving object data sets (Figure 5.10).

Figure 5.10: Particle system usage to show edge direction and density. Left original dataset, middle KDEEB (Hurter et al., 2012), and right ADEB (Peysakhovich et al., 2015). Directional graph visualization with a particle system. Particles can overlap but one can still visualize their direction.

The particle system uses the physique of particles with a lifetime, position, velocity, and acceleration (Blaas et al., 2009). These techniques help to show particle directions, and provide a synthetic view of global directions. The produced animations reveal the direction of airways, but also provide an interesting visual technique which is overlapping resistant. In certain conditions, one can still identify opposite direction flow even if they fully overlap (Figure 5.10).

In the following, we detail three use cases where particle systems help to better understand aircraft trajectory dataset. Aircraft follow flight routes—ordered sequences of geographically referenced locations. Actual aircraft trajectories do not always follow the exact location of the flight route. Air traffic controllers (ATCs) can dynamically alter routes for reasons of safety and optimization. Investigation of such flight routes, or traffic flows, is valuable to better understand airspace congestions and improve flight regulation and safety (Scheepens et al., 2015).

In this first-use case, we investigate two traffic flows crossing France in opposite directions. These flows correspond to flight transits between cities from the south-west to cities in the north-east, and vice versa: *Geneva*, Lyon, *Toulouse*, and *Madrid*. To avoid colliding aircraft, these parallel flows are geographically separated. Using our approach, we can select each flow with our selection widget, and define the altitude range corresponding to the desired flight routes—see Figure 5.11 where the altitude ranges are visualized using the small blue bars. The dynamics of these flows can be investigated separately, or can be directly compared by stacking their windows. As we can see, the traffic flow heading south-west is denser in the morning, while the traffic flow heading north-east is denser in the afternoon. According to our experts, this can be explained by passengers preferring to arrive in the morning in the south of Europe and wanting to return at the end of the day.

In this scenario, we investigate the aircraft distribution over different altitudes. In order to optimize fuel consumption, an aircraft remains at the same altitude, also called Flight Level (FL) as much as possible. FLs are expressed in feet—for instance, an aircraft flying at 30,000 feet (approximately 10 km high), has a FL of 300. A stabilized aircraft heading east (with a direction between 0 and 179°) have an odd FL (i.e., 310, 330, 350, 370, etc.). An aircraft heading west (with a direction between 180 and 359°) has an even FL (i.e., 300, 320, 340, 360, etc.). This mandatory rule helps to better separate aircraft traffic flows. When selecting flows by altitude, we can see the opposite aircraft direction between odd and even FL—see Figure 5.12. Since only powerful aircraft can reach high altitudes, the flow density decreases with higher FL.

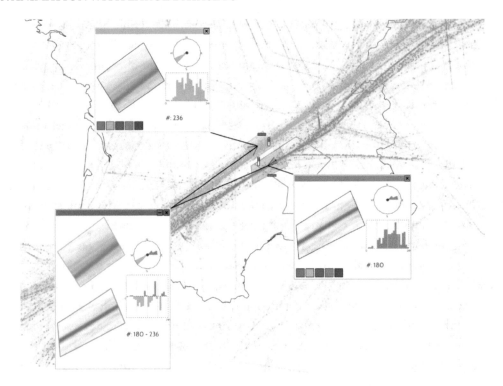

Figure 5.11: Traffic flows are visualized using a density map and a particle system. They can be selected using our novel selection widget, and are visible through colored particles. Two traffic flows of aircraft moving in opposite directions have been selected. The green traffic flow is moving southwest, and the orange traffic flow is heading north-east. Their dynamics can be explored and compared through movable windows, which show distributions of direction and density over time. The gray window shows the difference between the two traffic flows.

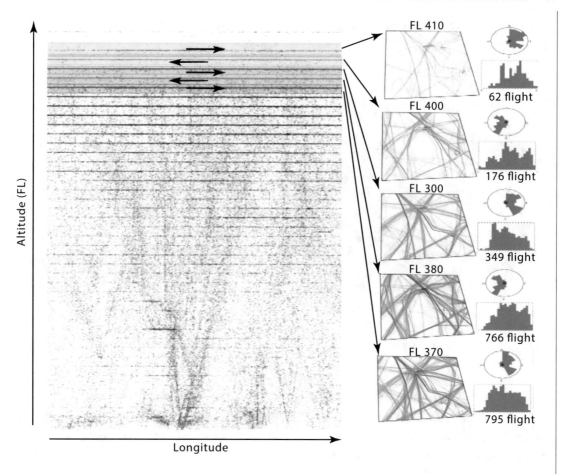

Figure 5.12: Vertical view (altitude, longitude). We can see aircraft flying east and aircraft flying west fly at alternating flight levels.

In this final use case, we investigate the traffic flow over Roissy Charles de Gaulle. When zooming into the Paris area, the flows appear entangled and spread out (Figure 5.13). A common approach, however, for dealing with such clutter is to apply a visual simplification technique such a edge bundling (see Section 4.5). Since flows are oriented, we use an extended version of the edge bundling technique which bundles trajectories with compatible directions (Peysakhovich et al., 2015). This technique reduces visual clutter by aggregating edges into bundled flows. Edge bundling provides a trade-off between empty spaces and overdrawing (Hurter et al., 2013a). As a drawback, the trajectories are distorted and thus not geographically accurate compared to the original trails. Figure 5.13 shows the double-cross flow system (four incoming flows and four outgoing), which can be investigated and compared. We can see some flows operate more in the morning, such as flow (1) heading east, while other flows operate mainly after midday, such as flow (2) coming from the south-east. Peaks correspond to hubs, i.e., when more aircraft arrive at the same time to maximize efficiency.

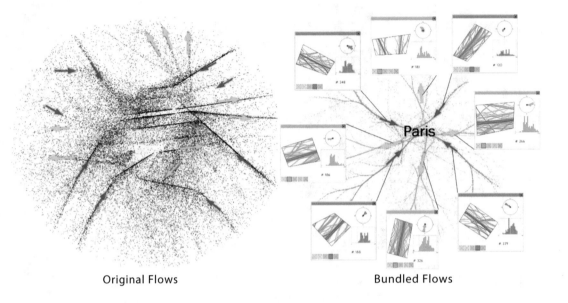

Original Flows   Bundled Flows

Figure 5.13: Left: An overview of traffic flows over the Paris area. Outgoing traffic flows have been marked with the green arrows, while incoming traffic flows have been marked with a red arrow. Right: The traffic flows have been bundled (Peysakhovichet al., 2015), selected, and the dynamics of these traffic flows are displayed using the movable windows.

5.2.1   PARTICLE SYSTEM REQUIREMENTS

To make such a particle system technique effective many interactions and technical challenges need to be addressed.

- Particle systems need a large amount of displayed entities which can hinder the interactive frame rate. Currently, without specific optimization a few million items can be displayed.

- New interactive tools must be implemented and validated, such as brushing, data deformation, and time navigation.

- Particle system assets must be clearly identified and validated with controlled experimentation.

Among the possible usages, investigation of eye tracking data is promising, but a design study must be performed beforehand to identify why and how particle systems can outperform existing visualization techniques. One of the major challenges is to prove that animation, which is an intrinsic part of a particle system, can speed-up the data retrieval process. One needs to watch the particle animation before being able to extract their direction. This question has already been investigated (Shanmugasundaram et al., 2007b) but not with this visualization technique.

5.3   DISTORTIONS

As previously explained, animation is an efficient tool to show the transition between two view configurations. Among the existing animations, the distortions show one with visual entities that get distorted along time. Considering visual entities like points, lines, polygons, or shapes, this distortion can be applied to a change of location but also to any visual variable (color, size, etc.; Bertin, 1983b). Carpendale et al. (1997) show the first instance of spatial distortion, Brosz et al. (2011) show efficient extension with maps, and Elmqvist et al. (2011) detail color distortion thanks to an interactive lens. Such animation can greatly help to understand how each visual entity gets distorted and helps to track a specific object.

In the following, we will detail a journey with simple distortions to more complex ones. Through this section, we will also show instances where distortions and animation can improve multidimensional data exploration.

5.3.1   2D LENS DISTORTION

Figure 5.14 shows the simple prototype using a lens to distort the view and thus remove occlusion. This prototype was inspired by the paper "Powers of Ten Thousand" (Lieberman, 1994) when navigating in large information space. The rationale was to use data deformation rather than data filtering and thus display all information in a distorted context. In this prototype, a lens pushes at its border points whose semantics do not correspond to the requested one (focus plus context technique). It also embeds labels with a basic avoidance process with a gradient computation (Tissoires, 2011). This prototype is CPU based and thus pretty slow (maximum 40,000 points could be handled with a reasonable frame rate).

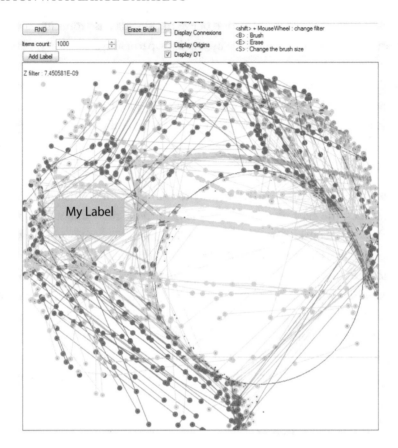

Figure 5.14: First mole view prototype with a semantic lens (Hurter et al., 2011).

Figure 5.15 explains the attraction and repulsion principle: the particle Pi is pushed away thanks to a repulsive vector field toward the border of the lens while it is also attracted toward its original position. This initial prototype computation is a vector force-based interactive system which resembles the dust and magnet system (Yi et al., 2005). The computation of the particle location is simply based on a vector addition. In order to avoid particle oscillation, each vector field must be weighed: the repulsive vector must be maximum when the particle is in the center of the lens, and must be null on its border (linear interpolation); the attraction vector must be null when the particle is at its original location, and maximum when its distance from its original location is more than the radius of the lens (Hurter et al., 2011).

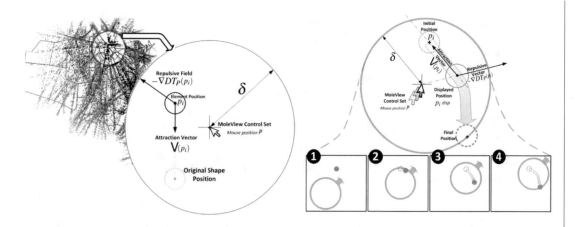

Figure 5.15: The moleview repulsion principle. The particle Pi is pushed away thanks to a repulsive vector field toward the border of the lens while it is also attracted toward its original position (Hurter et al., 2011).

5.3.2   3D LENS DISTORTION

The previous distortion shows smooth animation of 2D layout but no specific constrain can hinder the generalization to a 3D layout. The only constraint remains the size of the dataset and usually 3D visualization are larger datasets since they contain multilayered 2D layouts.

Figure 5.16 shows the first prototype which extended the mole view principle with a 3D layout with a sphere on which each voxel can be pushed to the edge of a semantic 3D lens. It also uses a point sprite visualization technique (Pajarola et al., 2004; Sainz and Pajarola, 2004). As such, each voxel (a 3D pixel) is displayed with a 2D point sprite whose border is transparent. This transparency follows a Gaussian transfer function. With a suitable transfer function one can visualize a more realistic picture such a 3D scan with a density value which corresponds to the density of the tissues: skin has a low density, bones a high density.

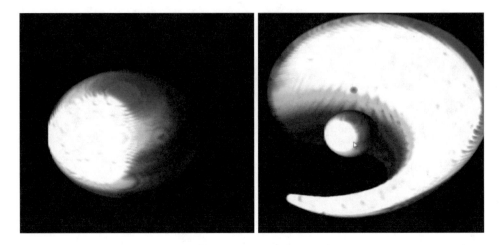

Figure 5.16: Point-based rendering of a 3D ball with a repulsive lens.

Figure 5.17 shows the first attempts to display 3D scan with a 2D layout followed by a 3D layout with pseudo color coding and finally a 3D layout with a transparent transfer function.

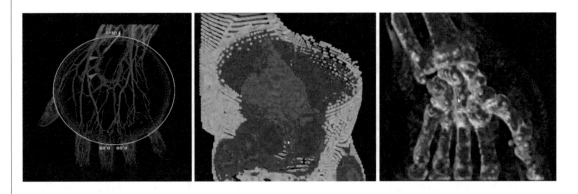

Figure 5.17: Lens distortion. Left: 2D image with the moleview which pushes away low-intensity pixels to reveal blood vessels. Center: 3D medical scan with a point-based rendering with a solid (no transparency) transfer function with a lens distortion to unveil the bones. Right: same visualization than the center image but with transparent colors.

This first version of the 3D mole view had many technical limitations (number of voxels, limited interactions, etc.) and thus a reduced contribution. Even if this implementation produced more aesthetic visualizations thanks to recent improvements with a powerful graphic card, previous works have already investigated such interactive techniques to explore such 3D visualization (McGuffin et al., 2003; Elmqvist, 2005). This prototype could not handle more than 400,000 voxels with a reasonable frame rate (10 fps). The bottleneck lies in the rendering process: the blending of point sprite is computationally challenging and the physics (movements around the 3D semantic lens) of the point sprite is also difficult. This prototype needs to update the position of these point sprites every frame to ensure a smooth animation.

Finally, color tunneling (Hurter et al., 2014c) details the solution to address this scalability issue with the render feedback buffer and the animation of a large multidimensional dataset. Next we show one instance of the distortion usage to show the skull from a medical scan.

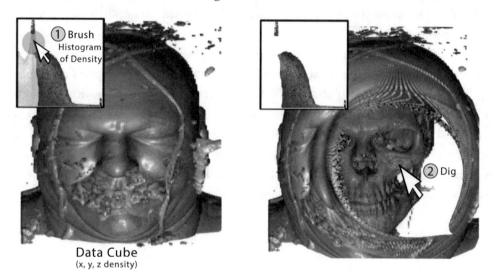

Figure 5.18: Color tunneling and the configurable lens to push away low-density voxels and reveal the skull.

Consider the 3D scans in Figure 5.5 and Figure 5.6 (128x128x112 voxels). One wants to "peek" inside the head to see the skull structure. The animation between the 3D scan and the gradient view show the brain and help to track it but not select it (Figure 5.5). For this, we first create a tissue-density histogram view and color its points by the DVR values given by gradient shading. The tall histogram peak indicates the largest voxel count in the volume, which are soft-tissue voxels. To the left of this peak, we see pink histogram points. These correspond to the skin tissue, which has the same color in the DVR view. In the middle of the peak, we see a thin dark vertical band. These are low-gradient voxels, which matches the fact that there are no density interfaces in soft tissue. In contrast, the pink histogram points are bright, which matches the DVR highlights at the skin-air and skin-soft tissue interfaces. Now that we understand the meaning of the histogram points, we select and lock all histogram points, and next select and unlock all (pink) points to the left of the peak. Finally, we apply the dig effect to the DVR view (Figure 5.18). This pushes unlocked voxels away from the focus, and reveals the hidden skull structure inside (gray).

5.3.3   BUNDLED DISTORTION

As a recap, the Figure 5.19 shows the different animations supported by the mole view (Hurter et al., 2011). The dual-layout shows an animation between two visual configurations (in the presented example, an image and its LCH color space visualization). The element-based exploration shows an animated distortion between a full set of trajectories and pushed trajectories which do not fulfill filtering requirement (altitude).

The bundled-based exploration is the last type of available distortion with the mole view. This distortion shows the animation between an original set of trajectory or a graph and its bundled version. This distortion can be applied to the full dataset or to the elements within the lens. This distortion can help track how visual entities evolve over time. This animation produces smooth animation and significantly helped to link edges from their original location to their bundled one. This investigation helped to detect flaws and parameter issues in the bundling process: too tight bends, incorrect bundles. This animation is also of great help for understanding the distortion and thus can be useful to setup, validate, or develop new deformation algorithms.

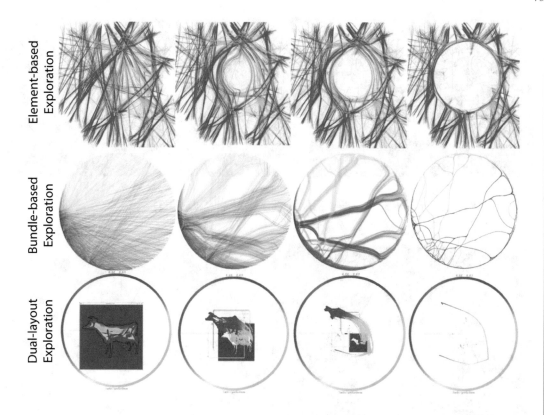

Figure 5.19: The three moleview principles. The animation between visual mapping (dual-layout exploration), the bundling-unbundling (bundle-based exploration), and the lens distortion (element-based exploration).

5.3.4 OBSTACLE AVOIDANCE

This distortion is not fully related to the previous one. The mole view already introduced a way to distort trajectory around free shape brushing area (Figure 5.20).

A formalized generalization of this principle is provided by Ersoy with the publication of the KDEB algorithm (Hurter et al., 2012). This technique, called obstacle avoidance, uses the distance transform to compute a way that an area will not be overlapped by bundled edges (Figure 5.21).

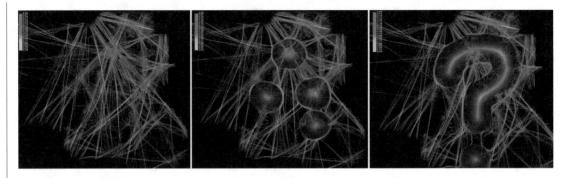

Figure 5.20: Brushing avoidance with the mole view (Hurter et al., 2011).

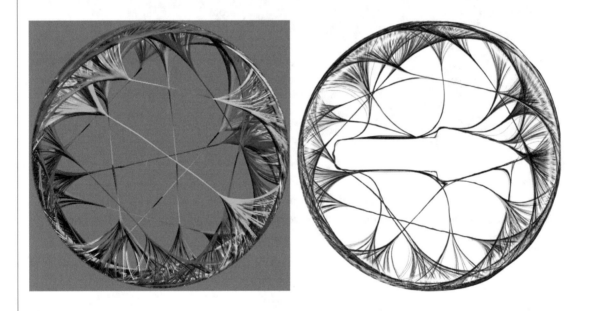

Figure 5.21: Bundling of dependency graph with obstacle ovoidance (right).

5.3.5   CASUAL INFOVIS: FREE DISTORTION, TRANSMOGRIFICATION

The last presented distortion is the project transmogrification (Brosz et al., 2013). This project was about developing a multitouch interactive system to deform visualization and thus leverage interactive data visualization and introduce the casual InfoVis. It is termed "casual" as the user is the one who transforms the view to find a better data visualization. As such, one can take the visualization of one's heartbeat (time series) and make it stick on its path displayed on Google map (Figure

5.22). This is where the casual InfoVis intervenes where the user has the power to transmogrify data representation to suit his or her will.

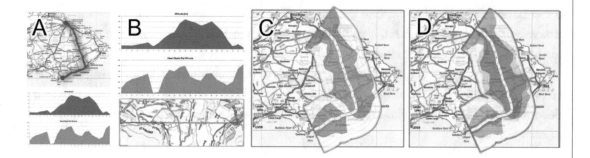

Figure 5.22: (A) Cycling data: a map with a route and area graphs with average altitude (purple) and heart-rate (orange) at each kilometer. (B) The rectified route map aligned under the linear graphs enables comparison of the measured variables to the map features. (C) The heart-rate graph wrapped around the route in the map shows effort in spatial context. (D) Same as C, but with multiple variables. Map used contains Ordnance Survey data © Crown.

This project uses the same animation principle provided by Color Tunneling (Hurter et al., 2014c): a linear interpolation between two visual layouts with a user-controlled animation (Figure 5.23).

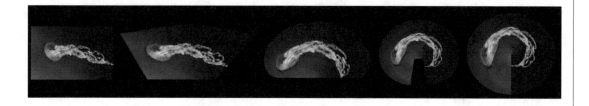

Figure 5.23: Linear interpolation between two shapes which create a simple morphing process.

5.4   CONCLUSION

This chapter explains how the simple idea of using every pixel as a discrete item can be of great interest to support multidimensional data exploration. Animations are featured alongside existing visualization software and are becoming a standard design feature (Bach et al., 2014a, 2014b) and are not limited to aesthetic design. Cordeil et al. (2013) managed to show how animation can display relevant information and identified three expected benefits of such animation: tracking graphical marks, understanding their relative arrangements, and perceiving structural elements.

Color tunneling (Hurter et al., 2014c) takes full advantage of animations and showed how it can be an efficient tool to dig into the dataset. Even if this chapter provides many instances of concrete data exploration success, it still lacks a proper evaluation to fully assess the power of animations to support data exploration.

Animation and distortion are at the crossroads of information visualization, visual analytics, computer graphics, and human-computer interaction (HCI) and much research remains to be done to unveil their full potential. Chapter 6 will detail some of the new research opportunities of such techniques.

CHAPTER 6

Research Outlook and Vision

As previously explained, image-based visualization techniques don't rely on only visualization but also on interaction. Their major challenge is to address the scalability issue and to do this the graphic card is a great help. The graphic cards and their GPU (graphical power unit) never stop improving in terms of computation power. The central processing unit (CPU) also does the same but the GPU and CPU differ in terms of their architecture and their goals. The CPU is the brain of the computer and it processes data while the GPU displays information. The CPU never stops improving its parallel processing power but it will never achieve the same performance and optimization as the GPU. GPU is devoted to processing massive amounts of data into parallel pixels (i.e., raster map). In this sense they are the best options to support image-based visualization.

To fully take advantage of such GPU power, many programming languages and specific Application Programming Interface (APIs) are available but none of them specialize in image-based visualization. This gap will be bridged shortly with extended and specific API derived from CUDA or OpenCL and will leverage many application domains thanks to image-based visualization.

This chapter will provide a very personal point of view about the research opportunities, pitfalls, and future challenges with image-based visualization. It will start with general statements and limitations with image-based processing. This will move onto the future challenges addressed by the previously detailed techniques: density map, edge bundling, and animation. Finally this chapter will conclude with future challenges and their uses for a few application domains.

6.1 GRAPHIC CARDS AND RASTER MAP

This section reports evidence on image-based usages. It will start with the example of the physics of light model for computer graphic rendering, followed by identified usages in data manipulation and exploration.

6.1.1 THE PHYSICS OF LIGHT IS A RENDERING PROCESS WITH MODERN GRAPHIC CARDS

SIGGRAPH is the premier international forum for disseminating new studies in computer graphics and interactive techniques. At this conference, many papers deal with the rendering process of a 3D scene with computer graphics. Among the rendering of technical challenges, the computation of illuminations and shadows is still an active investigation area. The ray casting (Roth, 1982) technique relies on the reverse propagation of light. This technique can take into account reflection, distraction, and absorption of material and produces realistic rendering. This is especially true with soft shadow computation—optical lens distortions which can be easily computed thanks to the physical modification of the light path. The ray casting algorithm is a simplified model of light propagation which omits the fact that every object in a 3D scene can reflect a given amount of light in every direction and thus be another light source. This simplification can prevent the illumination of an object when no direct light source is applied and thus produces areas without any shade. To address this issue more complex techniques have been developed such as the defining rendering equation (Kajiya, 1986) which uses light emission in many directions. This technique is far more computationally challenging compared to ray casting, even with some simplification and optimization (Purcell et al., 2002). Nevertheless, the rendering results are high quality images and are among the current techniques to produce theater movies.

However, these techniques have a drawback: their computation time prevents rendering in real time even with modern graphic cards. Therefore, other simplification models have been developed and especially ones that relies on rasterization and texturing. Textures are composed of juxtaposed colored pixels which form a rectangular shape. Texture can be applied to a 3D object and thus map specific shading or lightning features. These textures are the cornerstone of the simplified light rendering process. Instead of computing the complex physique of light, the rendering space is discretized and restricted to textures. A light model like Phong shading (Phong, 1975) is applied to vertexes (and then interpolated) or by pixels (fragment shader). Shades are computed by changing the point of view and then projecting the result on a texture (Engel, 2006). An environment map (Heidrich and Seidel, 1998) is computed and applied to a 3D object to simulate reflection. Many of these textures or pixel-based techniques are currently used to produce fast rendering in video games. The computation process consists of a succession of texture rendering and compositions with raster map processing (Pharr, 2005).

These raster usages produce a drastic simplification with much inaccurate or even incorrect rendering but are today essential rendering techniques. Fast, real-time rendering is achievable with rasterization techniques, and we still rely heavily on them.

6.1.2   DATA EXPLORATION AND MANIPULATION WITH IMAGE-BASED TECHNIQUES

Taking into account that computer displays are actual raster screens composed of juxtaposed color pixels, an image-based technique naturally fits their rendering process. Even if many image-based techniques do exist, few of them are used to perform data exploration. Many algorithms and interactive techniques are detailed with pedagogical approach in the book *Data Visualization* (Telea, 2014). Color tunneling (Hurter et al., 2014c) is one such system, where every piece of data is mapped to a discrete interactive item.

Recent advances in discretization and rasterization have introduced point cloud visualizations (Rusinkiewicz and Levoy, 2000). Rather than displaying texture surfaces, the point cloud technique uses each pixel as an individual object (i.e., color tunneling). These techniques arose with the development of 3D scanners and the management of large quantities of 3D pixels.

These techniques allow flexible data deformation and visualization. Data deformation can be performed with image-based techniques such as SBEB (Ersoy et al., 2011) and KDEEB (Hurter et al., 2012) to simplify graph and trail visualization. These recent techniques have also proven to be usable and scalable even if they rely on a significant data space simplification: data rasterization.

6.1.3   RASTER DATA INACCURACY

Raster data is less accurate than continuous forms; in computer graphics, floating values are more accurate than integers. Using this data simplification benefits computation time, but hinders accuracy. Now the question is to assess if this gain of performance is worth this data inaccuracy.

As previously explained, video games benefit from such a simplification and provide high visual quality video gaming with an interactive frame rate. If we take into account the KDEEB algorithm, small definition accumulation maps can perform bundling, but we can easily see these discretization artifacts (Figure 6.1).

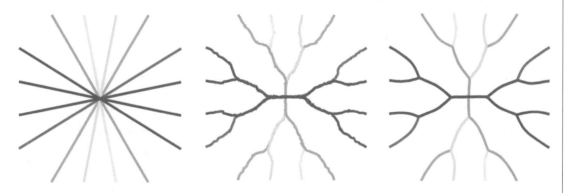

Figure 6.1: Raster map size effects. Original graph (left), bundled version with a small raster map (middle), and with a large raster map (right).

We also use very high-quality accumulation maps where the produced results are very smooth, but a lot of computation time is wasted by computing data from very similar locations (Figure 6.2). Parameter settings are open questions; today there is no easy way to compute the size of the accumulation map. Image-based bundling techniques show an interesting use case where drastically reduced accuracy (accumulation map of 100 pixels width) can produce exploitable bundling results. It is not strictly true that this algorithm is only pixel-based, since accumulation and data density are the solely pixel-based techniques, and once the computation of the density is completed, computation is performed in a double floating accuracy with the computation of the gradient value. Then the pixel-based rasterization is applied, since vertex will move according to the raster map, which also produces artifact, and then again the algorithm switches into double accuracy by smoothing the result. This example shows how the high-frequency data distortion can be reduced by using low-level filtering. To summarize, low accuracy produces a high-frequency artifact that can be easily filtered out and this is still worthwhile if computational speed is a requested feature.

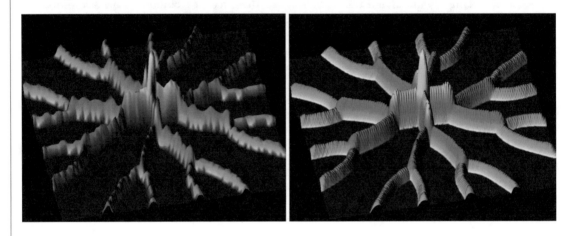

Figure 6.2: Small size density map (left), and large size density map (right).

6.2 FUTURE CHALLENGES

This book shows evidence of image-based techniques efficiency: they can provide a scalable solution and can also rely on technological advances such as faster and larger available memory for storage. The current bottleneck regarding GPGPU computation is the memory transfer between the GPU and the CPU, but we can forecast that future architecture improvement will address this issue.

In terms of technological improvements, one can expect a simpler way to program these powerful GPU processors, by providing high level API, and more advanced instruction set. Today, the major restriction is the lack of easy debugging tools. Such a GPU program, especially with shaders, needs special care to debug. It appends less frequently, but the system can lock. There is no easy way to investigate memory value and the debugging of high parallel threads is challenging even in a CPU and particularly so in the GPU.

Another interesting area of investigation lies in the use of massive memory and one can envisage memory as powerful as computation power. This investigation needs to be validated, but since memory is cheap and available in large quantity, it is a valuable resource. For instance, multithreaded computers are often in the idle state. We can take advantage of this waiting period to process information and store the result for future usages.

In the following we will especially detail opportunities with image-based visualization regarding edge bundling and animation/distortion. This book does not discuss density map future challenges since they are already mature enough and have already proved to be an efficient exploration tool with image-based visualization (Scheepens et al., 2012). As mentioned in Chapter 2, the computation time of a density map can still be improved to support even more interaction with larger density computation.

6.2.1   EDGE BUNDLING

This section will provide research investigation areas with image-based techniques. These extensions correspond to research questions, technical challenges which remain to be addressed. Bundling techniques are powerful tools for graph simplification. Bundlings have shown their ability to visually aggregate the links between nodes and thus produce empty spaces to improve a graph's global readability. This aggregation helps to retrieve information, like highlighting main flow and is often associated with interactive methods to improve this process.

Bundling techniques are useful for visual simplification of graphs and address issues regarding data density. Recent bundling techniques use graphic card power to support interactive visualization (OpenGL/WebGL/DirectX) or their ability to perform parallel computation (OpenCL/CUDA). The latest bundling techniques use image-based technique to reduce computation time with a complexity close to O(Edges).

The aesthetic criteria is undeniable with the production of smooth curves and the addition of image processing such as bump mapping (Figure 6.3). These aesthetic criteria have never been evaluated in terms of an asset to investigate datasets and it remains an open question.

Nevertheless, bundling techniques are not an all-purpose tool, and their ability to simplify visual graphs has many limitations. The following sections list bundling technique open questions that have not yet been investigated.

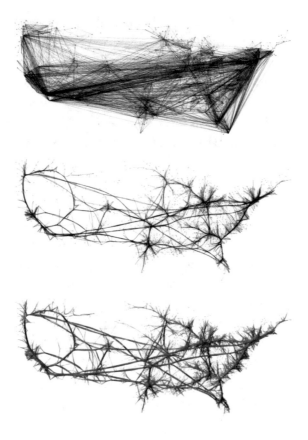

Figure 6.3: Bundling, original U.S. migration grap (top), KDEEB bundled version (middle), with bump mapping (bottom).

Dynamic Graphs

Dynamic graphs have not yet been intensely investigated. They need fast bundling computation to ensure that the bundle shows the up-to-date graph layout. Since a graph evolves over time, their bundled version must be updated without abrupt changes. Only KDEEB (Hurter et al., 2012) is able to offer a solution in this area. KDEEB has proved its effectiveness with a computation time of only a few milliseconds of calculation and ensures a continuous evolution of the graph layout thanks to the mean shift algorithm (Comaniciu and Meer, 2002). FDED (Holten and van Wijk, 2009) cannot ensure such continuity and the graph bundle can then show abrupt changes of structure (Nguyen et al., 2013a). Nevertheless, the lack of dynamic bundle exploitation tools remains and today there is no effective method provided. One can only watch the animation to interpret it.

Bundling Parameters

Bundling algorithms have shown that the bundled end result is highly dependent on the values of algorithm parameters. This is particularly the case with the pre-processing and clustering parameters that will strongly influence the final result. The bundled graph can thus be very highly or partially aggregated, and it is difficult to determine the correct settings without extensive testing.

Furthermore, bundling parameters are complex and linked to the algorithm. To some extent, these parameters should not be related to the bundling algorithm, but to the task the user wants to perform. For instance, one can adjust the ink ratio and the algorithm produces the corresponding bundling result.

Bundling Faithfulness and Accuracy

No previous work has investigated the accuracy of the displayed bundled information. As such, no metrics have been provided to assess the bundling algorithm quality. This limitation hinders their usages and the validation of new algorithms.

Finally, and most controversially, we can consider the faithfulness of the bundled results (Nguyen et al., 2013b). In this sense, there is no guarantee that the bundle version is representing information that is actually embedded in the data sets. For example, the processing of a random graph with KDEEB will highlight aggregates links. This can be partially explained because it is not possible to provide a uniform density graph link (blurring of the edges of the screen, Figure 6.4), but still shapes emerge.

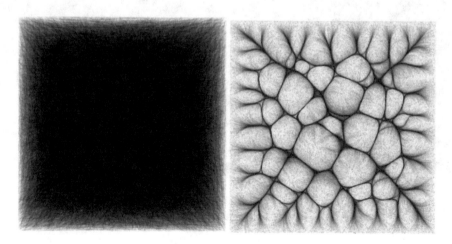

Figure 6.4: Graph pseudorandom and bundle version KDEEB (Hurter et al., 2012).

Even if patterns do emerge from the bundled view of the random graph in Figure 6.4, this does not necessarily mean that KDEEB is misleading. Since KDEEB is based on a gradient map which is linked to the resampled edge density, advection is highly sensitive to the variation of density. A theoretical random graph will have a uniform resampled edge density which is not the case in Figure 6.4 (each corner shows less nodes). This partially explains why patterns emerge after the bundling processing. One solution to address this issue would be to use a cycling wrapping view: edge can cross borders as if borders where connected in a spherical mapping. In this case the computed density map will be closed to a uniform map, the gradient close to zero and thus no edge advection will make patterns emerge.

Animated Bundling

Albeit visual scalability, current pixel-based techniques have not yet been widely applied to time-dependent datasets. Classic solutions, such as animation, work well for scientific datasets where the shapes to detect and track in time are natural and well known. However, we need to design new representations whose visual dynamics is both salient (so they can be easily detected) and suggestive (so their visual dynamics suggest to the user the desired types of events to be detected). For instance, in recent work, Hurter et al. (2013c) used animation to detect merging and splitting of groups when the shapes are bundle merges/splits, so these are easy to detect, and also suggestive (Figure 6.5).

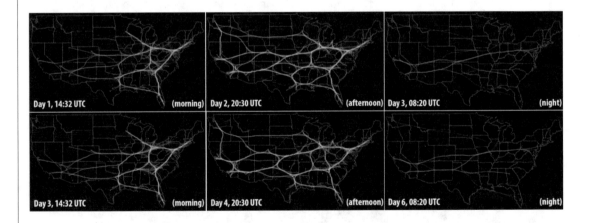

Figure 6.5: Small multiples of U.S. air lines over one week.

Movement data, which are multidimensional time-dependent data, describe changes of spatial positions of discrete mobile objects. Automatically collected movement data (e.g., GPS, RFID, radars, and others) are semantically poor as they basically consist of object identifiers, coordinates in space, and time stamps. Despite this, valuable information about the objects and their movement behavior as well as about the space and time in which they move can be gained from movement data by means of analysis. Analyzing and understanding time-dependent data poses additional non-trivial challenges to information visualization. First, such datasets are by their very nature several orders of magnitude larger than static datasets, which underlines the importance of relying on efficient interactions with multiple objects and fast algorithms. Second, while patterns of interest in static data can be naturally depicted by specific representations in still visualizations, we do not yet know how to best visualize dynamic patterns, which are inherent to time-dependent data. This type of data proposes fascinating perspectives with two research challenges to be addressed.

6.2.2   DISTORTION: POINT CLOUD DISPLAY

3D scanners are measurement instruments capable of collecting a large number of points in three-dimensional coordinates. To collect points, a laser beam is emitted in every direction and thanks to the beam-reflected time, the system can compute the distance of the targeted item. These points can be collected from any surface and from large or small size objects (e.g., a mechanical part of a few centimeters high or an archaeological site of several hectares). The accuracy of the measurement can vary from a few millimeters to a few centimeters depending on the distance and the targeted surface. These collected records form a point cloud and can be displayed. This new source of information, in addition to those already used for topological statements (e.g., GPS), open new fields of application with accurate data in the areas of civil engineering and infrastructure, architecture, construction, and preservation of cultural heritage. Academic research and professional usages of these data have already started but they are mostly limited to the 3D modeling process of the scanned surfaces. Today, 3D scanners are more accessible in terms of price and usability; as such, an increasing number of point clouds are available for potential usages.

Point clouds are composed of discrete elements and for this reason image-based techniques can be applied to process this type of dataset. In the conventional approaches to 3D modeling, computing power is dedicated to the 3D reconstruction performed by the machine. Thanks to image-based techniques we can adopt an innovative approach using the computing power not for the automatic modeling but to provide interactive data manipulation (Figure 6.6). This approach offers the advantage of giving more flexibility to the 3D modeling process. In other words, the computing power of machines will still be used, but it will be the operator, with interactive tools, that will guide the data processing. Building such a tool is challenging since a large data set must be displayed and manipulated in real time. Color tunneling was developed for this purpose but with limited interactions (digging into the dataset). Point cloud visualization and manipulation is an interesting application domain to investigate and will provide a way to assess how interactivity can potentially outperform automatic methods, and thus study the optimal amount of automatic as opposed to manual algorithm to provide in order to support efficient 3D modeling.

Figure 6.6: 3D point scan (left), with edge filtering (right, MoleView) (Hurter et al., 2011).

6.3   IMAGE-BASED ALGORITHM IN APPLICATION DOMAINS

This section details project perspectives where the use of image-based algorithms may support specific activities. This will provide the opportunity to answer research questions with respect to concrete examples (internal validity). For instance, the exploration and visualization of a large point cloud remains a technical challenge. Air traffic analysis is also an inexhaustible source of inspiration. A detailed description of image-based visualization is provided for air traffic activities (Hurter et al., 2014b). For instance, wind parameters can be extracted from aircraft trajectories (Hurter et al., 2014a).

This research area is built upon technological improvements and advances in graphic card powers and is directly linked to the Ben Shneiderman mantra "overview first, zoom and filter, then derails on demand" (Shneiderman, 1996). The ultimate goal is to provide interactive scalable tools that are built upon image based algorithms. This image-based concept has been previously introduced (McDonnel and Elmqvist, 2009) as well as big data challenges (Fekete and Plaisant, 2002). Many previous works tried to address scalability issue with data reduction methods, resampling (Das Sarma et al., 2012), filtering (Ahlberg and Shneiderman, 1994), and aggregation (Carr et al., 1987). Interactive systems have also been especially developed to investigate these scalability issues (Liu et al., 2013).

Today visualization of a large dataset is still a challenge and the bottleneck lies in the available pixels on the screen which are not numerous enough to visualize every piece of information in large datasets. Therefore, aggregation techniques to simplify the visualization of trails and graphs with edge bundling techniques are of great interest. In the same way, interactive techniques to manipulate and to link data, smooth animations and continuous space interactions, are the cornerstone of seamless interactive data exploration techniques and decision making. Image-based algorithms provide a wide area of investigation where much still remains to be done.

6.3.1   EYE TRACKING

Eye tracking systems help to capture user gaze and are becoming a very popular data source to analyze users' activity. Eye trackers record lots of information which need to be processed and thus is a major challenge due to the data size. A state-of-the-art of existing technique is provided in the EuroVis Star report (Blascheck et al., 2014). Image based techniques are already available to address analyzing issue with dynamic bundled visualization (Hurter et al., 2013a) and with recorded image around the gaze point (Kurzhals et al., 2015) but lots of unexplored investigation remains to be done.

As an example, the CReA (research center of the French army) investigates pilot training with such eye tracking systems. During their training, pilots need to learn specific gaze scanning patterns, such as looking outside to prevent collision with other aircraft, then looking inside the cockpit to gather and analyze flight parameters (altitude, speed, direction, engine setting). While this pattern is simple, it remains an issue for pilots beginning their training who spend most of their time looking inside the cockpit. This behavior is especially true with simulator training (Johnson et al., 2006) and it must be corrected as fast as possible to reduce negative transfer from ground-simulation to real flight. Therefore, an experiment has been developed to improve ground-simulation teaching. In this experimentation (Figure 6.7), the pilot is notified with different modalities (audio or visual alarms) when he or she has an incorrect ocular behavior (e.g., looking for more than 2s inside the cockpit). Depending on the notification modality, the time to learn a specific skill is expected to be different. Hence, one will seek the most appropriate notification. For instance, one notification will hide cockpit information, and the pilot will have to look outside to make this information reappear. Another modality will be to emit a warning audio message, and then the user will decide to look back outside or not.

Figure 6.7: Simulation environment with head mounted eye tracking system (left). Visualization of fixation points inside the cockpit when gathering flight parameters in red, and elsewhere in blue.

Figure 6.8: Gaze analysis with bundling technique (Hurter et al., 2013a). Color corresponds to the gaze direction. Trail width corresponds to the density of the path.

Image-based visualization can be used at different levels to handle user gaze: data recording, data analysis (Dubois et al., 2015), and gaze interaction.

- **Data recording:** Head-mounted eye tracking systems do not always provide third-party software to capture in real-time user gaze (only post processing) (Imbert et al., 2015). Furthermore, calibration issues can occur over time (mainly data shift). Due to the data size, image-based techniques can help to address these two issues and are promising investigation areas thanks to image processing with GP-GPU techniques.

- **Data analysis:** Collected data (gaze location and fixation time) must be processed to investigate user behavior. As such, bundling methods help to simplify eye trails and thus help their analysis (Figure 6.8) but many over investigation remains to be done.

- **Gaze interaction:** While image-based techniques are not the most common technique to perform gaze interaction (Traoré et al., 2015), they are great opportunity to build more responsive systems.

6.3.2 IMAGE PROCESSING: SKIN CANCER INVESTIGATION

The diagnosis and prognosis of skin tumors, such as melanoma, is an increasingly important health issue. Given the growing prevalence of such tumors, designing efficient and effective methods to analyze imaged skin lesions is an important goal, e.g., for the automatic measurement of the ABCD (asymmetry, border, colors, and dimension) diagnostic score.

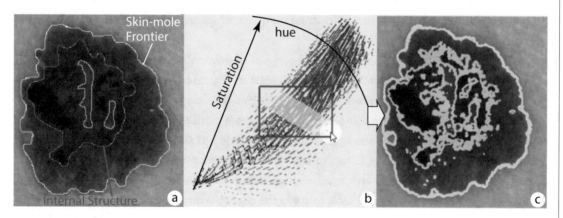

Figure 6.9: Skin tumor investigation with color tunneling. One can navigate between data configuration and brush pixel to select a skin-mole frontier (Hurter et al., 2014c).

Color tunneling (Hurter et al., 2014c) already investigated image processing techniques with the manipulation of skin tumor images. Figure 6.9 shows how this software helps to segment the image with a pixel-based technique. The user can navigate between view configurations, brush specific area, and select skin-mole frontier. These initial results will be extended with additional interaction techniques such the ABCD score evolution over time.

6.3.3   COGNITIVE MAPS AND ALZHEIMER DISEASE

According to a ministerial review of 2004, approximately 860,000 people are affected by Alzheimer's disease in France. This estimation will possibly reach 1.3 million in 2020 and 2.1 million in 2040. This is why the study of Alzheimer's disease has been identified as a major society challenge. In the PAQUID cohort study, 3,777 elderly people performed a lexical evocation task by orally producing the most possible city names within 3 min. They were longitudinally followed during a 22-yr period and some of them developed the dementia of Alzheimer (Rondeau et al., 2009; Tessier et al., 2001; Nejjari et al., 1997). Among the different tasks and questionnaires reported in this study, we will especially investigate the lexical evocation task. This task is directly related to the concept of the cognitive map introduced in 1948 by Edward Tolman in his article "Cognitive Maps in Rats and Men" and echoed by O'Keefe and Nadel (1978) in their book *The Hippocampus as a Cognitive Map*. The cognitive map is considered, among other things, as a sort of matrix in which the episodes of life can be stored. Formally, the cognitive map is "the capacity of a subject to reorganize spatial information in order to develop cognitive representations of the environment beyond its field of perception." It locates mentally several geographical points against each other, and organizes them into a coherent configuration. The analysis of the production of city names provides a unique opportunity to study the spatial mental representation of French geographical space for elderly people before and after developing the dementia.

The following reports on an unexpected usage of edge bundling technique with the memory project and the Alzheimer disease. This project started in September 2014 with the collaboration of neuroscientists and computer graphics experts. We can visualize this cognitive map by connecting cited cities with a line. Since this visualization becomes cluttered with the numerous lines, we applied the KDEEB bundling technique (Figure 6.10). One can also compare the maps produced at different time periods. The first investigations show a stronger graph implication with an elderly person suffering from dementia. This first investigations need to be validated. The objective of this project is to develop techniques and tools to study the cognitive map of elderly subjects in the years preceding the onset of Alzheimer's disease. The developed tools will use image-based technique (e.g., bundling) and will help to better understand the processes involved in the building and deterioration of cognitive maps.

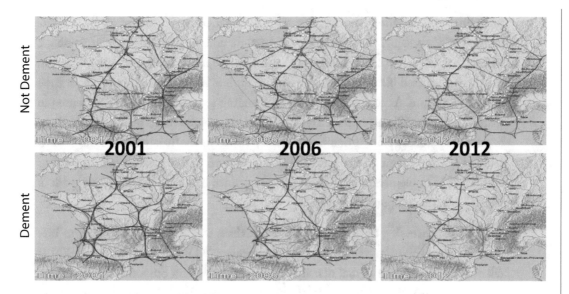

Figure 6.10: Visualization of cited cities over time. We notice a significant simplification of the network for demented subjects.

6.4   CONCLUSION

This book summarizes investigations with GPU usages to support data visualization and manipulation. The following lists the expended benefits of such techniques and I will describe their limitations and future challenges.

- **Benefits from the graphic cards and their massive memory and parallel computation power usage:** FromDaDy (Hurter et al., 2009b) uses a GPU brushing technique, the MoleView (Hurter et al., 2011) uses animation techniques, and color tunneling (Hurter et al., 2014c) deals with larger datasets. Recent advances in graphic card power introduced dedicated programming languages which are not restrained to graphic visualization but to general computation (compute shader, CUDA, OpenCL, etc.). These new languages will not replace existing rendering pipeline since we still need to display images, and shaders propose an efficient and mature transformation process. However, these new languages provide interesting additional features which are complementary to existing visualization processing. They can be combined to existing rendering pipeline and thus leverage interactive visualization computation. For instance, 3D models can be produced with standard programmable pipeline while physics of these objects (inertia, accelerations, etc.) can be computed with compute shaders. Regarding information visualization, the exact same process can occur; color tunneling takes advantage of the programmable rendering pipeline to produce point based rendering, while the points' movements are computed with a render feedback buffer (former version of the compute shader). In the future, this kind of architecture may be generalized but it has a major constraint: it is more complex in terms of program architecture. As such it is far more challenging to produce a GPGPU technique than a standard CPU algorithm. The GPUGPU technique often needs to be refined to support parallel computation and memory concurrency (it is valid to read and to write at the same time in the same memory location). In conclusion, before starting a GPUGPU implementation, one needs to consider if the extra amount of work is worth the extra computational power.

- **The image processing field offers many algorithms that are worth applying to image-based information visualization.** For instance, KDEEB (Hurter et al., 2012) takes advantages of the kernel density estimation algorithm. Smooth animated bundles use the mean shift algorithm (Comaniciu and Meer, 2002) (clustering algorithm). Data densities can be displayed by simple lighting computation. By synthesizing color, shading, and texture at a pixel level, one can achieve a much higher freedom in constructing a wide variety of representation that is able to depict the rich data patterns we aim to analyze. Many image-processing algorithms are available and can be tested with data exploration technique. The goal is to provide efficient tools in terms of data exploration, interaction and visualization. The expected benefits rely on shorter computation processing time while extracting relevant information. Since these image-processing algorithms operate on "pixels" or discrete data space, we need to identify if this large data inaccuracy does not spoil data exploration. As an example, KDEEB bundling turned out to be the fastest bundling algorithm with valid bundling results. Even if DKEEB (Hurter et al., 2013a) uses a discrete accumulation map with a reduced number of cells, the visual bundling results are still valid. In the transformation process, the density map helps to compute a gradient map which is not a discreet data space any more. As an open question, we do not yet know the validity of these data simplifications and when such a process can be applied.

- **The use of memory instead of computation can reduce algorithm complexity:** Benefits have already been explored in many projects. FromDaDy uses a texture to store trail identifiers and to drastically reduce brushing computation, the MoleView uses a gradient map (stored in the GPU memory) to distort trails or pixels, and KDEEB uses the same gradient map to compute the bundled version of a graph. It is obvious that storing computation results can speed up the algorithm since it is faster to read a memory cell than to perform multiple computations. However, this expected benefit does not only rely on data storage, but also on algorithm complexity. As such, KDEEB transforms the initial bundling complexity from $O(E.(E-1))$ to $O(E)$ where E are edges of a graph. This algorithm simplification works thanks to the second expected benefit (applying image-processing algorithm; in this case the mean shift clustering), but it also introduces a new data-processing pipeline where memory storage plays the central role. This new pipeline may, under given circumstances, simplify algorithm implementations. This last expected benefit has not yet been deeply investigated and this is more a research question than an actual fact and it remains a promising investigation area.

As our knowledge of GPGPU and image-based algorithms improves, it also opens up new opportunities for visualization tool research. Such data exploration will become faster, more interactive, and will handle larger datasets. As a final conclusion, image-based InfoVis will definitely improve data visualization and interaction to support efficient decision making and this opens up a large new investigation field for researchers.

Bibliography

Ahlberg, C. and Shneiderman, B., 1994. Visual Information Seeking: Tight Coupling of Dynamic Query Filters with Starfield Displays, in: *Proceedings of the SIGCHI Conference on Human Factors in Computing Systems, CHI '94*. ACM, New York, NY, pp. 313–317. DOI: 10.1145/191666.191775. 88

Amini, F., Henry Riche, N., Lee, B., Hurter, C., and Irani, P., 2015. Understanding Data Videos: Looking at Narrative Visualization Through the Cinematography Lens, in: *Proceedings of the 33rd Annual ACM Conference on Human Factors in Computing Systems, CHI '15*. ACM, New York, NY, pp. 1459–1468. DOI: 10.1145/2702123.2702431. 1

Andrienko, G. and Andrienko, N., 2008. Spatio-temporal aggregation for visual analysis of movements, in: *IEEE Symposium on Visual Analytics Science and Technology, 2008. VAST '08*. Presented at the IEEE Symposium on Visual Analytics Science and Technology, 2008. VAST '08, pp. 51–58. DOI: 10.1109/VAST.2008.4677356. 64

Appert, C., Beaudouin-Lafon, M., and Mackay, W.E., 2005. Context matters: Evaluating Interaction Techniques with the CIS Model, in: FSEDA, S.F.B., MA, LHG, MSc, P.M.Ms., Moore, D., MBCS, R.R.Bs., , CEng (Eds.), *People and Computers XVIII — Design for Life*. Springer London, pp. 279–295. 11

Archambault, D., Purchase, H.C., and Pinaud, B., 2011. Animation, Small Multiples, and the Effect of Mental Map Preservation in Dynamic Graphs. *IEEE Trans. Vis. Comput. Graph.* 17, 539–552. DOI: 10.1109/TVCG.2010.78. 55

Auber, D., 2004. Tulip — A Huge Graph Visualization Framework, in: Jünger, M., and Mutzel, P. (Eds.), *Graph Drawing Software, Mathematics and Visualization*. Springer Berlin Heidelberg, pp. 105–126. DOI: 10.1007/978-3-642-18638-7_5. 7

Auber, D., Chiricota, Y., Jourdan, F., and Melançon, G., 2003. Multiscale Visualization of Small World Networks, in: *Proceedings of the Ninth Annual IEEE Conference on Information Visualization, INFOVIS'03*. IEEE Computer Society, Washington, DC, USA, pp. 75–81. DOI: 10.1109/infvis.2003.1249011.

Bach, B., Pietriga, E., and Fekete, J.-D., 2014a. Visualizing dynamic networks with matrix cubes. ACM Press, pp. 877–886. DOI: 10.1145/2556288.2557010. 77

Bach, B., Pietriga, E., and Fekete, J.-D., 2014b. GraphDiaries: Animated Transitions andTemporal Navigation for Dynamic Networks. *IEEE Trans. Vis. Comput. Graph.* 20, 740–754. DOI: 10.1109/TVCG.2013.254. 77

Baudel, T., 2004. Browsing Through an Information Visualization Design Space, in: *CHI '04 Extended Abstracts on Human Factors in Computing Systems, CHI EA '04*. ACM, New York, NY, USA, pp. 765–766. DOI: 10.1145/985921.985925. 17

Baudisch, P., Tan, D., Collomb, M., Robbins, D., Hinckley, K., Agrawala, M., Zhao, S., and Ramos, G., 2006. Phosphor: Explaining Transitions in the User Interface Using Afterglow Effects, in: *Proceedings of the 19th Annual ACM Symposium on User Interface Software and Technology, UIST '06*. ACM, New York, NY, USA, pp. 169–178. DOI: 10.1145/1166253.1166280. 15

Bertin, J., 1983a. *Semiology of Graphics*. University of Wisconsin Press, Madison, WI. 17,

Bertin, J., 1983b. *Semiology of Graphics*. University of Wisconsin Press, Madison, WI. 69

Bezerianos, A., Chevalier, F., Dragicevic, P., Elmqvist, N., and Fekete, J.D., 2010. Graphdice: A System for Exploring Multivariate Social Networks, in: *Proceedings of the 12th Eurographics / IEEE - VGTC Conference on Visualization, EuroVis'10*. Eurographics Association, Aire-la-Ville, Switzerland, pp. 863–872. DOI: 10.1111/j.1467-8659.2009.01687.x. 8, 55, 56

Bezerianos, A., Dragicevic, P., and Balakrishnan, R., 2006. Mnemonic Rendering: An Image-based Approach for Exposing Hidden Changes in Dynamic Displays, in: *Proceedings of the 19th Annual ACM Symposium on User Interface Software and Technology, UIST '06*. ACM, New York, NY, pp. 159–168. DOI: 10.1145/1166253.1166279. 15

Blaas, J., Botha, C., Grundy, E., Jones, M., Laramee, R., and Post, F., 2009. Smooth Graphs for Visual Exploration of Higher-Order State Transitions. *IEEE Trans. Vis. Comput. Graph.* 15, 969–976. DOI: 10.1109/TVCG.2009.181. 64, 65

Blascheck, T., Kurzhals, K., Raschke, M., Burch, M., Weiskopf, D., and Ertl, T., 2014. State-of-the-art of visualization for eye tracking data. *Proc. EuroVis* 2014. 89

Brosz, J., Carpendale, S., and Nacenta, M.A., 2011. The Undistort Lens, in: *Proceedings of the 13th Eurographics / IEEE - VGTC Conference on Visualization, EuroVis'11*. The Eurographs Association and John Wiley & Sons, Ltd., Chichester, UK, pp. 881–890. DOI: 10.1111/j.1467-8659.2011.01937.x. 69

Brosz, J., Nacenta, M.A., Pusch, R., Carpendale, S., and Hurter, C., 2013. Transmogrification: Causal Manipulation of Visualizations, in: *Proceedings of the 26th Annual ACM Sympo-*

sium on User Interface Software and Technology, UIST '13. ACM, New York, NY, USA, pp. 97–106. DOI: 10.1145/2501988.2502046. 76

Card, S.K., and Mackinlay, J., 1996. *The Structure of the Information Visualization Design Space*. 12. 15, 16, 17

Card, S.K., Mackinlay, J.D., and Shneiderman, B. (Eds.), 1999a. *Readings in Information Visualization: Using Vision to Think*. Morgan Kaufmann Publishers Inc., San Francisco, CA. 2, 4, 5

Card, S.K., Mackinlay, J.D., and Shneiderman, B. (Eds.), 1999b. *Readings in Information Visualization: Using Vision to Think*. Morgan Kaufmann Publishers Inc., San Francisco, CA. 11, 12

Card, S.K., Newell, A., and Moran, T.P., 1983. *The Psychology of Human-Computer Interaction*. L. Erlbaum Associates Inc., Hillsdale, NJ. 11

Card, S.K., Robertson, G.G., and Mackinlay, J.D., 1991. The Information Visualizer, an Information Workspace, in: *Proceedings of the SIGCHI Conference on Human Factors in Computing Systems, CHI '91*. ACM, New York, NY, pp. 181–186. DOI: 10.1145/108844.108874. 55

Carpendale, M.S.T., Cowperthwaite, D.J., and Fracchia, F.D., 1997. Extending Distortion Viewing from 2D to 3D. *IEEE Comput. Graph. Appl.* 17, 42–51. DOI: 10.1109/38.595268. 69

Carr, D.B., Littlefield, R.J., Nicholson, W.L., and Littlefield, J.S., 1987. Scatterplot Matrix Techniques for Large N. *J. Am. Statist. Assoc.* 82, 424–436. DOI: 10.2307/2289444. 88

Chevalier, F., Dragicevic, P., and Hurter, C., 2012. Histomages: Fully Synchronized Views for Image Editing, in: *Proceedings of the 25th Annual ACM Symposium on User Interface Software and Technology, UIST '12*. ACM, New York, NY, pp. 281–286. DOI: 10.1145/2380116.2380152. 57, 58

Cleveland, W.S., 1993. *Visualizing Data*. Hobart Press, Summit, NJ. 23

Cohen, M., Puech, C., Sillion, F., Haeberli, P., and Segal, M., 1993. *Texture Mapping as a Fundamental Drawing Primitive*. 4, 8

Comaniciu, D. and Meer, P., 2002. Mean shift: a robust approach toward feature space analysis. *IEEE Trans. Pattern Anal. Mach. Intell.* 24, 603–619. DOI: 10.1109/34.1000236. 2, 43, 84, 94

Conversy, S., Chatty, S., and Hurter, C., 2011. Visual Scanning As a Reference Framework for Interactive Representation Design. *Inf. Vis.* 10, 196–211. DOI: 10.1177/1473871611415988. 17

Cordeil, M., Hurter, C., Conversy, S., and Causse, M., 2013. Assessing and Improving 3D Rotation Transition in Dense Visualizations, in: *Proceedings of the 27th International BCS Human*

Computer Interaction Conference, BCS-HCI '13. British Computer Society, Swinton, UK, UK, pp. 7:1–7:10. 56, 77

CUDA Technology, 2007. http://www.nvidia.com/cuda. 6

Cui, W., Zhou, H., Qu, H., Wong, P.C., and Li, X., 2008. Geometry-Based Edge Clustering for Graph Visualization. *IEEE Trans. Vis. Comput. Graph.* 14, 1277–1284. DOI: 10.1109/TVCG.2008.135. 39, 40

Das Sarma, A., Lee, H., Gonzalez, H., Madhavan, J., and Halevy, A., 2012. Efficient Spatial Sampling of Large Geographical Tables, in: *Proceedings of the 2012 ACM SIGMOD International Conference on Management of Data, SIGMOD '12.* ACM, New York, NY, pp. 193–204. DOI: 10.1145/2213836.2213859. 88

Davis, T.A. and Hu, Y., 2011. The University of Florida Sparse Matrix Collection. *ACM Trans Math Softw* 38, 1:1–1:25. DOI: 10.1145/2049662.2049663. 51

Dickerson, M., Eppstein, D., Goodrich, M.T., and Meng, J.Y., 2003. Confluent Drawings: Visualizing Non-planar Diagrams in a Planar Way, in: Liotta, G. (Ed.), *Graph Drawing, Lecture Notes in Computer Science.* Springer Berlin Heidelberg, pp. 1–12. 39

Dubois, E., Blättler, C., Camachon, C., and Hurter, C., 2015. Eye Movements Data Processing for Ab Initio Military Pilot Training, in: *Intelligent Decision Technologies.* Springer International Publishing,Berlin Heidelberg. pp. 125–135. DOI: 10.1007/978-3-319-19857-6_12. 90

Dwyer, T., Marriott, K., and Wybrow, M., 2007. Integrating Edge Routing into Force-Directed Layout, in: Kaufmann, M., Wagner, D. (Eds.), *Graph Drawing, Lecture Notes in Computer Science.* Springer, Berlin Heidelberg, pp. 8–19. DOI: /10.1007/978-3-540-70904-6_3. 39

Ellis, G. and Dix, A., 2007. A Taxonomy of Clutter Reduction for Information Visualisation. *IEEE Trans. Vis. Comput. Graph.* 13, 1216–1223. DOI: 10.1109/TVCG.2007.70535. 39

Elmqvist, N., 2005. BalloonProbe: Reducing Occlusion in 3D Using Interactive Space Distortion, in: *Proceedings of the ACM Symposium on Virtual Reality Software and Technology, VRST '05.* ACM, New York, NY, pp. 134–137. DOI: 10.1145/1101616.1101643. 72

Elmqvist, N., Dragicevic, P., and Fekete, J., 2011. Color Lens: Adaptive Color Scale Optimization for Visual Exploration. *IEEE Trans. Vis. Comput. Graph.* 17, 795–807. DOI: 10.1109/TVCG.2010.94. 69

Elmqvist, N., Dragicevic, P., and Fekete, J., 2008. Rolling the Dice: Multidimensional Visual Exploration using Scatterplot Matrix Navigation. *IEEE Trans. Vis. Comput. Graph.* 14, 1539–1148. DOI: 10.1109/TVCG.2008.153. 8, 55, 56

Engel, W., 2006. *Shader X4 Advanced Rendering Techniques*, Har/Cdr edition. ed. Charles River Media, Hingham, MA. 80

Ersoy, O., Hurter, C., Paulovich, F., Cantareiro, G., and Telea, A., 2011. Skeleton-Based Edge Bundling for Graph Visualization. *IEEE Trans. Vis. Comput. Graph.* 17, 2364–2373. DOI: 10.1109/TVCG.2011.233. 40, 41, 42, 51, 81

Fekete, J.-D. and Plaisant, C., 2002. Interactive Information Visualization of a Million Items, in: *Proceedings of the IEEE Symposium on Information Visualization (InfoVis'02), INFOVIS '02.* IEEE Computer Society, Washington, DC, p. 117–124. DOI: 10.1109/infvis.2002.1173156. 1, 8, 88

Fitzmaurice, G., Matejka, J., Mordatch, I., Khan, A., and Kurtenbach, G., 2008. *Safe 3D Navigation.* ACM Press, New York, p. 7. DOI: 10.1145/1342250.1342252. 56

Frishman, Y. and Tal, A., 2007. Multi-Level Graph Layout on the GPU. *IEEE Trans. Vis. Comput. Graph.* 13, 1310–1319. DOI: 10.1109/TVCG.2007.70580. 7

Gansner, E.R., Hu, Y., North, S., and Scheidegger, C., 2011. Multilevel agglomerative edge bundling for visualizing large graphs, in: *2011 IEEE Pacific Visualization Symposium (PacificVis).* Presented at the 2011 IEEE Pacific Visualization Symposium (PacificVis), pp. 187–194. DOI: 10.1109/PACIFICVIS.2011.5742389. 39, 40, 51

Gansner, E.R. and Koren, Y., 2007. Improved Circular Layouts, in: Kaufmann, M., Wagner, D. (Eds.), *Graph Drawing, Lecture Notes in Computer Science.* Springer Berlin Heidelberg, pp. 386–398. DOI: 10.1007/978-3-540-70904-6_37. 39

Hadlak, S., Schulz, H., and Schumann, H., 2011. In Situ Exploration of Large Dynamic Networks. *IEEE Trans. Vis. Comput. Graph.* 17, 2334–2343. DOI: 10.1109/TVCG.2011.213. 39

Harris, M., 2005. Mapping Computational Concepts to GPUs, in: *ACM SIGGRAPH 2005 Courses, SIGGRAPH '05.* ACM, New York, NY. DOI: 10.1145/1198555.1198768. 26

He, B., Fang, W., Luo, Q., Govindaraju, N.K., and Wang, T., 2008. Mars: A MapReduce Framework on Graphics Processors, in: *Proceedings of the 17th International Conference on Parallel Architectures and Compilation Techniques, PACT'08.* ACM, New York, NY, pp. 260–269. DOI: 10.1145/1454115.1454152. 6

Heer, J. and Robertson, G.G., 2007. Animated Transitions in Statistical Data Graphics. *IEEE Trans. Vis. Comput. Graph.* 13, 1240–1247. DOI: 10.1109/TVCG.2007.70539. 55, 56

Heidrich, W. and Seidel, H.-P., 1998. View-independent Environment Maps, in: *Proceedings of the ACM SIGGRAPH/EUROGRAPHICS Workshop on Graphics Hardware, HWWS '98.* ACM, New York, NY, p. 39–ff. DOI: 10.1145/285305.285310. 80

Henry, N., Fekete, J., and McGuffin, M.J., 2007. NodeTrix: a Hybrid Visualization of So-
    cial Networks. *IEEE Trans. Vis. Comput. Graph.* 13, 1302–1309. DOI: 10.1109/
    TVCG.2007.70582. 39

Holten, D., 2006. Hierarchical Edge Bundles: Visualization of Adjacency Relations in Hierarchical
    Data. *IEEE Trans. Vis. Comput. Graph.* 12, 741–748. DOI: 10.1109/TVCG.2006.147. 7,
    39, 40, 42, 43, 51

Holten, D. and van Wijk, J.J., 2009. Force-directed Edge Bundling for Graph Visualization, in:
    *Proceedings of the 11th Eurographics/IEEE - VGTC Conference on Visualization, Eu-
    roVis'09.* Eurographics Association, Aire-la-Ville, Switzerland, pp. 983–998. DOI:
    10.1111/j.1467-8659.2009.01450.x. 40, 42, 51, 56, 84

Hurter, C., 2014. *Image Based Algorithms to Support Interactive Data Exploration (Habilitation à
    Direger des Recherches).* Université Paul Sabatier Toulouse, France. 2

Hurter, C., 2010. *Caractérisation de Visualisations et Exploration Interactive de Grandes Quantités de
    Données Multidimensionnelles.* Université Paul Sabatier - Toulouse, France III. 11, 12

Hurter, C., Alligier, R., Gianazza, D., Puechmorel, S., Andrienko, G., and Andrienko, N., 2014a.
    Wind Parameters Extraction from Aircraft Trajectories. *Comput. Environ. Urban Syst.,
    Progress in Movement Analysis – Experiences with Real Data* 47, 28–43. DOI: 10.1016/j.
    compenvurbsys.2014.01.005. 88

Hurter, C. and Conversy, S., 2008. Towards Characterizing Visualizations, in: Graham, T.C.N.,
    Palanque, P. (Eds.), *Interactive Systems. Design, Specification, and Verification, Lecture Notes
    in Computer Science.* Springer, Berlin Heidelberg, pp. 287–293. DOI: 10.1007/978-3-
    540-70569-7_26. 12, 16, 17

Hurter, C., Conversy, S., Gianazza, D., and Telea, A.C., 2014b. Interactive Image-based Informa-
    tion Visualization for Aircraft Trajectory Analysis. *Transp. Res. Part C Emerg. Technol.*
    DOI: 10.1016/j.trc.2014.03.005. 33, 40, 88

Hurter, C., Conversy, S., and Vinot, J.-L., 2009a. Temporal Data Visualizations for Air Traffic
    Controllers (ATC), in: *Interacting with Temporal Data, CHI 2009 Workshop.* 17

Hurter, C., Ersoy, O., Fabrikant, S., Klein, T., and Telea, A., 2013a. Bundled Visualization of
    Dynamic Graph and Trail Data. *IEEE Trans. Vis. Comput. Graph.* Early Access Online.
    DOI: :10.1109/TVCG.2013.246. 7, 9, 44, 67, 89, 90, 94

Hurter, C., Ersoy, O., and Telea, A., 2013b. Smooth Bundling of Large Streaming and Sequence
    Graphs, in: *Visualization Symposium (PacificVis), 2013 IEEE Pacific.* Presented at the
    Visualization Symposium (PacificVis), 2013 IEEE Pacific, pp. 41–48. DOI: 10.1109/
    PacificVis.2013.6596126. 39, 40

Hurter, C., Ersoy, O., and Telea, A., 2013c. Smooth Bundling of Large Streaming and Sequence Graphs, in: *Visualization Symposium (PacificVis), 2013 IEEE Pacific*. Presented at the Visualization Symposium (PacificVis), 2013 IEEE Pacific, pp. 41–48. DOI: 10.1109/PacificVis.2013.6596126. 42, 43, 86

Hurter, C., Ersoy, O., and Telea, A., 2012. Graph Bundling by Kernel Density Estimation. *Comp Graph Forum* 31, 865–874. John Wiley & Sons, Inc. DOI: 10.1111/j.1467-8659.2012.03079.x. 2, 3, 7, 9, 42, 43, 44, 45, 47, 51, 64, 75, 81, 84, 85, 94

Hurter, C., Kapp, V., and Conversy, S., 2008. An Infovis Approach to Compare ATC Comets, in: *ICRAT 2008, International Conference on Research in Air Transportation*. 16

Hurter, C., Taylor, R., Carpendale, S., and Telea, A., 2014c. Color Tunneling: Interactive Exploration and Selection in Volumetric Datasets, in: *2014 IEEE Pacific Visualization Symposium (PacificVis)*. Presented at the 2014 IEEE Pacific Visualization Symposium (PacificVis), pp. 225–232. DOI: 10.1109/PacificVis.2014.61. 29, 59, 62, 73, 77, 78, 80, 91, 93

Hurter, C., Telea, A., and Ersoy, O., 2011. MoleView: An Attribute and Structure-Based Semantic Lens for Large Element-Based Plots. *IEEE Trans. Vis. Comput. Graph.* 17, 2600–2609. DOI: 10.1109/TVCG.2011.223. 7, 40, 53, 56, 57, 58, 70, 71, 74, 76, 88, 93

Hurter, C., Tissoires, B., and Conversy, S., 2010. Accumulation as a Tool for Efficient Visualization of Geographical and Temporal Data, in: *AGILE Workshop Geospatial Visual Analytics: Focus on Time*. 23

Hurter, C., Tissoires, B., and Conversy, S., 2009b. FromDaDy: Spreading Aircraft Trajectories Across Views to Support Iterative Queries. *IEEE Trans. Vis. Comput. Graph.* 15, 1017–1024. DOI: 10.1109/TVCG.2009.145. 7, 8, 19, 23, 27, 40, 55, 56, 93

Hurter, C., Tissoires, B., and Conversy, S., 2009. FromDaDy: Spreading Aircraft Trajectories Across Views to Support Iterative Queries. *IEEE Trans. Vis. Comput. Graph.* 15, 1017–1024. DOI: 10.1109/TVCG.2009.145.

Imbert, J.-P., Hurter, C., Peysakhovich, V., Blättler, C., Dehais, F., and Camachon, C., 2015. Design Requirements to Integrate Eye Trackers in Simulation Environments: Aeronautical Use Case, in: *Intelligent Decision Technologies*. Springer International Publishing, Berlin, Heidelberg, pp. 231–241. DOI: 10.1007/978-3-319-19857-6_21. 90

Jankun-Kelly, T.J., Dwyer, T., Holten, D., Hurter, C., Nöllenburg, M., Weaver, C., and Xu, K., 2014. Scalability Considerations for Multivariate Graph Visualization, in: Kerren, A., Purchase, H.C., Ward, M.O. (Eds.), *Multivariate Network Visualization, Lecture Notes in Computer Science*. Springer International Publishing, Berlin, Heidelberg, pp. 207–235. DOI: 10.1007/978-3-319-06793-3_10. 5

Johnson, N., Wiegmann, D., and Wickens, C., 2006. Effects of Advanced Cockpit Displays on General Aviation Pilots' Decisions to Continue Visual Flight Rules Flight into Instrument Meteorological Conditions. *Proc. Hum. Factors Ergon. Soc. Annu. Meet.* 50, 30–34. DOI: 10.1177/154193120605000107. 89

Kajiya, J.T., 1986. The Rendering Equation, in: *Proceedings of the 13th Annual Conference on Computer Graphics and Interactive Techniques, SIGGRAPH '86*. ACM, New York, NY, pp. 143–150. DOI: 10.1145/15922.15902. 80

Koffa, K., 1963. *Principles of Gestalt Psychology*. Harcourt, Brace & World, New York. 14

Kurzhals, K., Hlawatsch, M., Heimerl, F., Burch, M., Ertl, T., and Weiskopf, D., 2015. Gaze Stripes: Image-Based Visualization of Eye Tracking Data. *IEEE Trans. Vis. Comput. Graph.* PP, 1–1. DOI: 10.1109/TVCG.2015.2468091. 89

Lambert, A., Bourqui, R., and Auber, D., 2010a. Winding Roads: Routing Edges into Bundles, in: *Proceedings of the 12th Eurographics / IEEE - VGTC Conference on Visualization, EuroVis'10*. Eurographics Association, Aire-la-Ville, Switzerland, pp. 853–862. DOI: 10.1111/j.1467-8659.2009.01700.x. 7, 40

Lambert, A., Bourqui, R., and Auber, D., 2010b. Winding Roads: Routing Edges into Bundles, in: *Proceedings of the 12th Eurographics / IEEE - VGTC Conference on Visualization, EuroVis'10*. The Eurographs Association and John Wiley & Sons, Ltd., Chichester, UK, pp. 853–862. DOI: 10.1111/j.1467-8659.2009.01700.x. 39

Lambert, A., Bourqui, R., and Auber, D., 2010c. 3D Edge Bundling for Geographical Data Visualization, in: *Information Visualisation (IV), 2010 14th International Conference*. Presented at the Information Visualisation (IV), 2010 14th International Conference, pp. 329–335. DOI: 10.1109/IV.2010.53. 39

Lee, B., Parr, C.S., Plaisant, C., Bederson, B.B., Veksler, V.D., Gray, W.D., and Kotfila, C., 2006a. TreePlus: Interactive Exploration of Networks with Enhanced Tree Layouts. *IEEE Trans. Vis. Comput. Graph.* 12, 1414–1426. DOI: 10.1109/TVCG.2006.106. 39

Lee, B., Plaisant, C., Parr, C.S., Fekete, J.-D., and Henry, N., 2006b. Task Taxonomy for Graph Visualization, in: *Proceedings of the 2006 AVI Workshop on BEyond Time and Errors: Novel Evaluation Methods for Information Visualization, BELIV '06*. ACM, New York, NY, pp. 1–5. DOI: 10.1145/1168149.1168168. 53

Lieberman, H., 1994. Powers of Ten Thousand: Navigating in Large Information Spaces, in: *Proceedings of the 7th Annual ACM Symposium on User Interface Software and Technology, UIST '94*. ACM, New York, NY, pp. 15–16. DOI: 10.1145/192426.192434. 69

Liu, Z., 2012. *Network-based Visual Analysis of Tabular Data*. Georgia Institute of Technology, Atlanta, GA. 39

Liu, Z., Jiang, B., and Heer, J., 2013. imMens: Real-time Visual Querying of Big Data, in: *Proceedings of the 15th Eurographics Conference on Visualization, EuroVis '13*. The Eurographs Association and John Wiley & Sons, Ltd., Chichester, UK, pp. 421–430. DOI: 10.1111/cgf.12129. 88

McDonnel, B. and Elmqvist, N., 2009. Towards utilizing GPUs in information visualization: A model and implementation of image-space operations. *IEEE Trans. Vis. Comput. Graph.* 15, 1105–1112. DOI: 10.1109/TVCG.2009.191. 4, 7, 8

McDonnel, B. and Elmqvist, N., 2009. Towards Utilizing GPUs in Information Visualization: A Model and Implementation of Image-space Operations. *IEEE Trans. Vis. Comput. Graph.* 15, 1105–1112. DOI: 10.1109/TVCG.2009.191. 88

McGuffin, M.J., Tancau, L., and Balakrishnan, R., 2003. Using Deformations for Browsing Volumetric Data, in: *IEEE Visualization, 2003*. VIS 2003. Presented at the IEEE Visualization, 2003. VIS 2003, pp. 401–408. DOI: 10.1109/VISUAL.2003.1250400. 72

Munshi, A., Gaster, B., Mattson, T.G., Fung, J., and Ginsburg, D., 2011. *OpenCL Programming Guide*, 1st ed. Addison-Wesley Professional. http://www.amazon.fr/OpenCL-Programming-Guide-Aaftab-Munshi/dp/0321749642. 6

Nejjari, C., Tessier, J.F., Baldi, I., Barberger-Gateau, P., Dartigues, J.F., and Salamon, R., 1997. [Epidemiologic aspects of respiratory aging: contribution of the PAQUID survey]. Rev. Dépidémiologie Santé Publique 45, 417–428. 92

Nguyen, Q., Eades, P., and Hong, S.-H., 2013a. StreamEB: Stream Edge Bundling, in: *Proceedings of the 20th International Conference on Graph Drawing, GD'12*. Springer-Verlag, Berlin, Heidelberg, pp. 400–413. DOI: 10.1007/978-3-642-36763-2_36. 43, 84

Nguyen, Q., Eades, P., and Hong, S.-H., 2013b. On the Faithfulness of Graph Visualizations, in: *Proceedings of the 20th International Conference on Graph Drawing, GD'12*. Springer-Verlag, Berlin, Heidelberg, pp. 566–568. DOI: 10.1007/978-3-642-36763-2_55. 85

O'Keefe, J. and Nadel, L., 1978. *The Hippocampus as a Cognitive Map*. Clarendon Press Oxford.

Owens, J.D., Houston, M., Luebke, D., Green, S., Stone, J.E., and Phillips, J.C., 2008. GPU Computing. *Proc. IEEE 96*, 879–899. DOI: 10.1109/JPROC.2008.917757. 19

Owens, J.D., Luebke, D., Govindaraju, N., Harris, M., Krüger, J., Lefohn, A.E., and Purcell, T., 2007. A *Survey of General-purpose Computation on Graphics Hardware*. Comp. Graph. For. 26-1, pp. 80–113. http://onlinelibrary.wiley.com/doi/10.1111/j.1467-8659.2007.01012.x/abstract. 5, 7

Pajarola, R., Sainz, M., and Guidotti, P., 2004. Confetti: object-space point blending and splatting. *IEEE Trans. Vis. Comput. Graph.* 10, 598–608. DOI: 10.1109/TVCG.2004.19. 71

Peysakhovich, V., Hurter, C., and Telea, A., 2015. Attribute-Driven Edge Bundling for General Graphs with Applications in Trail Analysis. *PacificVis.* DOI: 10.1109/pacificvis.2015.7156354. 45, 47, 49, 50, 64, 67, 68

Phan, D., Xiao, L., Yeh, R., Hanrahan, P., and Winograd, T., 2005. Flow Map Layout, in: *Proceedings of the Proceedings of the 2005 IEEE Symposium on Information Visualization, INFOVIS '05.* IEEE Computer Society, Washington, DC, p. 219–224. DOI: 10.1109/INFOVIS.2005.13. 39

Pharr, M., 2005. *GPU Gems 2: Programming Techniques for High-Performance Graphics and General-Purpose Computation*, 1st ed. Addison-Wesley Professional, Upper Saddle River, NJ. 80

Phong, B.T., 1975. Illumination for Computer Generated Pictures. *Commun ACM* 18, 311–317. DOI: 10.1145/360825.360839. 30, 80

Purcell, T.J., Buck, I., Mark, W.R., and Hanrahan, P., 2002. Ray Tracing on Programmable Graphics Hardware, in: *Proceedings of the 29th Annual Conference on Computer Graphics and Interactive Techniques, SIGGRAPH '02.* ACM, New York, NY, USA, pp. 703–712. DOI: 10.1145/566570.566640. 80

Qu, H., Zhou, H., and Wu, Y., 2007. Controllable and Progressive Edge Clustering for Large Networks, in: Kaufmann, M., Wagner, D. (Eds.), *Graph Drawing, Lecture Notes in Computer Science.* Springer, Berlin Heidelberg, pp. 399–404. DOI: 10.1007/978-3-540-70904-6_38. 39

Rekimoto, J., 1997. Pick-and-drop: A Direct Manipulation Technique for Multiple Computer Environments, in: *Proceedings of the 10th Annual ACM Symposium on User Interface Software and Technology, UIST '97.* ACM, New York, NY, pp. 31–39. DOI: 10.1145/263407.263505. 27

Renso, C., Spaccapietra, S., and Zimányi, E., 2013. *Mobility Data.* Cambridge University Press, Cambridge, UK. DOI: 10.1017/CBO9781139128926. 34

Rondeau, V., Jacqmin-Gadda, H., Commenges, D., Helmer, C., and Dartigues, J.-F., 2009. Aluminum and Silica in Drinking Water and the Risk of Alzheimer's Disease or Cognitive Decline: Findings from 15-Year Follow-up of the PAQUID Cohort. *Am. J. Epidemiol.* 169, 489–496. DOI: 10.1093/aje/kwn348. 92

Roth, S.D., 1982. Ray casting for modeling solids. *Comput. Graph. Image Process.* 18, 109–144. DOI: 10.1016/0146-664X(82)90169-1. 79

Rusinkiewicz, S. and Levoy, M., 2000. QSplat: A Multiresolution Point Rendering System for Large Meshes, in: *Proceedings of the 27th Annual Conference on Computer Graphics and Interactive Techniques, SIGGRAPH '00*. ACM Press/Addison-Wesley Publishing Co., New York, NY, pp. 343–352. DOI: 10.1145/344779.344940. 81

Sainz, M. and Pajarola, R., 2004. Point-based Rendering Techniques. *Comput Graph* 28, 869–879. DOI: 10.1016/j.cag.2004.08.014. 71

Scheepens, R., Hurter, C., van de Wetering, H., and van Wijk, J.J., 2015. Visualization, Selection, and Analysis of Traffic Flows. *IEEE Trans. Vis. Comput. Graph*. p. 1–1. DOI: 10.1109/TVCG.2015.2467112. 65

Scheepens, R., Willems, N., Van de Wetering, H., Andrienko, G., Andrienko, N., and van Wijk, J.J., 2011. Composite Density Maps for Multivariate Trajectories. *IEEE Trans. Vis. Comput. Graph*. 17, 2518–2527. DOI: 10.1109/TVCG.2011.181. 7, 24, 40

Scheepens, R., Willems, N., van de Wetering, H., and van Wijk, J., 2012. Interactive Density Maps for Moving Objects. *IEEE Comput. Graph. Appl*. 32, 56–66. DOI: 10.1109/MCG.2011.88. 83

Schulz, H.-J. and Hurter, C., 2013. Grooming the Hairball - How to Tidy up Network Visualizations? Presented at the INFOVIS 2013 tutorial, IEEE Information Visualization Conference, Atlanta, GA. 9

Segel, E., Heer, J., 2010. Narrative Visualization: Telling Stories with Data. *IEEE Trans. Vis. Comput. Graph*. 16, 1139–1148. DOI: 10.1109/TVCG.2010.179. 1

Selassie, D., Heller, B., and Heer, J., 2011. Divided Edge Bundling for Directional Network Data. *IEEE Trans. Vis. Comput. Graph*. 17, 2354–2363. DOI: 10.1109/TVCG.2011.190. 40

Shanmugasundaram, M., Irani, P., and Gutwin, C., 2007a. Can Smooth View Transitions Facilitate Perceptual Constancy in Node-link Diagrams? ACM Press, New York, p. 71. DOI: 10.1145/1268517.1268531. 55

Shanmugasundaram, M., Irani, P., and Gutwin, C., 2007b. Can Smooth View Transitions Facilitate Perceptual Constancy in Node-link Diagrams?, in: *Proceedings of Graphics Interface 2007, GI '07*. ACM, New York, NY, pp. 71–78. DOI: 10.1145/1268517.1268531. 69

Shneiderman, B., 1996. The eyes have it: A task by data type taxonomy for information visualizations, in: *IEEE Symposium on Visual Languages*. pp. 336–343. DOI: 10.1109/vl.1996.545307. 88

Shreiner, D. and Group, T.K.O.A.W., 2009. *OpenGL Programming Guide: The Official Guide to Learning OpenGL*, Versions 3.0 and 3.1, 7th ed. Addison-Wesley Professional. 25, 29

Silverman, B.W., 1986. *Density Estimation for Statistics and Data Analysis*. Chapman & Hall, London. DOI: 10.1007/978-1-4899-3324-9. 23, 24

Sutherland, I., 2012. The Tyranny of the Clock. *Commun ACM* 55, 35–36. DOI: 10.1145/2347736.2347749. 6

Taylor, A.R., Gibson, S.J., Peracaula, M., Martin, P.G., Landecker, T.L., Brunt, C.M., Dewdney, P.E., Dougherty, S.M., Gray, A.D., Higgs, L.A., Kerton, C.R., Knee, L.B.G., Kothes, R., Purton, C.R., Uyaniker, B., Wallace, B.J., Willis, A.G., and Durand, D., 2003. The Canadian Galactic Plane Survey. *Astron. J.* 125, 3145. DOI: 10.1086/375301. 62, 63

Telea, A., 2007. *Data Visualization: Principles and Practice*. A K Peters/CRC Press, Wellesley, MA.

Telea, A. 2014. *Data Visualisation*. https://www.crcpress.com/Data-Visualization-Principles-and-Practice-Second-Edition/Telea/9781466585263. 80

Telea, A. and Ersoy, O., 2010. Image-based Edge Bundles: Simplified Visualization of Large Graphs, in: *Proceedings of the 12th Eurographics / IEEE - VGTC Conference on Visualization, EuroVis'10*. Eurographics Association, Aire-la-Ville, Switzerland, pp. 843–852. DOI: 10.1111/j.1467-8659.2009.01680.x. 40, 41

Telea, R. and Wijk, J.J.V., 2002. An Augmented Fast Marching Method for Computing Skeletons and Centerlines, in: In *Proc. of the Symposium on Data Visualisation (VisSym'02)*, 2002. pp. 251–259. 40

Tessier, J.F., Nejjari, C., Letenneur, L., Filleul, L., Marty, M.L., Gateau, P.B., and Dartigues, J.F., 2001. Dyspnea and 8-year Mortality among Elderly Men and Women: The PAQUID Cohort Study. *Eur. J. Epidemiol.* 17, 223–229. DOI: 10.1023/A:1017977715073. 92

Thomas, J.J. and Cook, K.A., 2005. *Illuminating the Path: The Research and Development Agenda for Visual Analytics*. IEEE Computer Society Press, Washington, DC. 1

Thompson, C.J., Hahn, S., and Oskin, M., 2002. Using Modern Graphics Architectures for General-purpose Computing: A Framework and Analysis, in: *35th Annual IEEE/ACM International Symposium on Microarchitecture, 2002. (MICRO-35)*. Proceedings. Presented at the 35th Annual IEEE/ACM International Symposium on Microarchitecture. (MICRO-35). Proceedings, pp. 306–317. DOI: 10.1109/MICRO.2002.1176259. 6

Thrower, N.J.W., 1969. Edmond Halley as a Thematic Geo-Cartographer. *Ann. Assoc. Am. Geogr.* 59, 652–676. DOI: 10.1111/j.1467-8306.1969.tb01805.x. 13

Tissoires, B., 2011. Conception (Instrumenté), Réalisation et Optimisation de Graphismes Interactifs (Ph.D.). Université de Toulouse, Université Toulouse III - Paul Sabatier, Toulouse, France. 19, 69

Traoré, M., Marmisse, L., Zimmerman, C., André, M., and Hurter, C., 2015. e-ATC: Using an Eye-tracker to Instrument the Activity of Air Traffic Control. ACM Press, New York, pp. 1–10. DOI: 10.1145/2820619.2820631. 91

Tufte, E.R., 1986. *The Visual Display of Quantitative Information*. Graphics Press, Cheshire, CT. 51

Tversky, B., Morrison, J.B., and Betrancourt, M., 2002. Animation: Can It Facilitate? *Int. J. Hum.-Comput. Stud.* 57, 247–262. DOI: 10.1006/ijhc.2002.1017. 55

Van Ham, F. and Wattenberg, M., 2008. Centrality Based Visualization of Small World Graphs, in: *Proceedings of the 10th Joint Eurographics / IEEE - VGTC Conference on Visualization, EuroVis'08*. Eurographics Association, Aire-la-Ville, Switzerland, pp. 975–982. DOI: 10.1111/j.1467-8659.2008.01232.x. 39

Van Liere, R. and Leeuw, W. de, 2003. GraphSplatting: Visualizing Graphs as Continuous Fields. *IEEE Trans. Vis. Comput. Graph.* 9, 206–212. DOI: 10.1109/TVCG.2003.1196007. 24, 40

Van Wijk, J.J. and Telea, A., 2001. Enridged Contour Maps, in: *Proceedings of the Conference on Visualization '01, VIS '01*. IEEE Computer Society, Washington, DC, pp. 69–74. DOI: 10.1109/visual.2001.964495. 24

Van Wijk, J.J. and van de Wetering, H., 1999. Cushion Treemaps: Visualization of Hierarchical Information, in: *Proceedings of the 1999 IEEE Symposium on Information Visualization, INFOVIS '99*. IEEE Computer Society, Washington, DC, pp. 73–78, 147. DOI: 10.1109/INFVIS.1999.801860. 8

Von Landesberger, T., Kuijper, A., Schreck, T., Kohlhammer, J., van Wijk, J. j., Fekete, J.-D., and Fellner, D. W., 2011. Visual Analysis of Large Graphs: State-of-the-Art and Future Research Challenges. *Comput. Graph. Forum* 30, 1719–1749. DOI: 10.1111/j.1467-8659.2011.01898.x. 9

Wilkinson, L., Wills, D., Rope, D., Norton, A., Dubbs, R., 2005. *The Grammar of Graphics*, 2nd edition. ed. Springer, New York. 17

Willems, N., van de Wetering, H., and van Wijk, J.J., 2009. Visualization of Vessel Movements, in: *Proceedings of the 11th Eurographics / IEEE - VGTC Conference on Visualization, EuroVis'09*. Eurographics Association, Aire-la-Ville, Switzerland, pp. 959–966. DOI: 10.1111/j.1467-8659.2009.01440.x. 24

Wise, J.A., Thomas, J.J., Pennock, K., Lantrip, D., Pottier, M., Schur, A., and Crow, V., 1995. Visualizing the Non-visual: Spatial Analysis and Interaction with Information from Text Documents, in: *Proceedings of the 1995 IEEE Symposium on Information Visualization,*

INFOVIS '95. IEEE Computer Society, Washington, DC, p. 51. DOI: 10.1109/infvis.1995.528686. 24

Yi, J.S., Melton, R., Stasko, J., and Jacko, J.A., 2005. Dust & Magnet: Multivariate Information Visualization Using a Magnet Metaphor. *Inf. Vis.* 4, 239–256. DOI: 10.1057/palgrave.ivs.9500099. 70

Young, D. and Shneiderman, B., 1993. A Graphical Filter/Flow Representation of Boolean Queries: A Prototype Implementation and Evaluation. *J. Am. Soc. Inf. Sci.* 44, 327–339. DOI: 10.1002/(SICI)1097-4571(199307)44:6<327::AID-ASI3>3.0.CO;2-J. 20

Zhou, H., Yuan, X., Cui, W., Qu, H., and Chen, B., 2008. Energy-Based Hierarchical Edge Clustering of Graphs, in: *Visualization Symposium, 2008. PacificVIS '08*. IEEE Pacific. Presented at the Visualization Symposium, 2008. PacificVIS '08. IEEE Pacific, pp. 55–61. DOI: 10.1109/PACIFICVIS.2008.4475459. 39

Author Biography

Christophe Hurter is a professor at the Interactive Computing Laboratory (LII) of the French Civil Aviation University (ENAC) in Toulouse, France. In 2010 he received his Ph.D. in computer science from the University of Toulouse and in 2014 he received his Habilitation à Diriger des Recherche. Christophe is also an associate researcher at the research center for the French military Air Force (CReA), Salon de Provence, France. His research interests include information visualization (InfoVis) and human-computer interaction (HCI), particularly the visualization of multivariate data in space and time. He also investigates the design of scalable visual interfaces and the development of image-based rendering techniques. Throughout his career he has worked on several projects, including large data exploration tools, graph simplification (edge bundling), and paper-based interaction.

Printed in the United States
by Baker & Taylor Publisher Services